Levels of Genetic Control in Development

The Thirty-Ninth Symposium of
The Society for Developmental Biology
Storrs, Connecticut, June 23–25, 1980

Executive Committee
1979–1980

Ursula K. Abbott, University of California at Davis, President
Norman K. Wessells, Stanford University, Past-President
Paul B. Green, Stanford University, President-Designate
W. Sue Badman, CMBD, NIGMS, NIH, Secretary
John G. Scandalios, North Carolina State University, Treasurer
Gerald M. Kidder, University of Western Ontario, Member-at-Large

1980–1981

Paul B. Green, Stanford University, President
Ursula K. Abbott, University of California at Davis, Past-President
Fotis C. Kafatos, Harvard University, President-Designate
W. Sue Badman, CMBD, NIGMS, NIH, Secretary
John G. Scandalios, North Carolina State University, Treasurer
Gerald M. Kidder, University of Western Ontario, Member-at-Large

Business Manager
Holly Schauer
P.O. Box 40741
Washington, D.C. 20016

Levels of Genetic Control in Development

Stephen Subtelny, Editor

Department of Biology
Rice University
Houston, Texas

Ursula K. Abbott, Coeditor

Department of Avian Sciences
University of California
Davis, California

Alan R. Liss, Inc. • New York

Address all Inquiries to the Publisher
Alan R. Liss, Inc., 150 Fifth Avenue, New York, NY 10011

Copyright © 1981 Alan R. Liss, Inc.

Printed in the United States of America.

Library of Congress Cataloging in Publication Data
Main entry under title:

Levels of genetic control in development.

(Symposium of the Society for Developmental Biology; 39th)
Includes bibliographical references and index.
1. Developmental genetics — Congresses. 2. Genetic regulation — Congresses. I. Subtelny, Stephen Stanley, 1925–   . II. Abbott, Ursula K. III. Series: Symposium (Society for Developmental Biology); 39th.
QH511.S6 39th [QH453]   574.3s [574.3]   81-1952
ISNB 0-8451-1500-6   AACR2

Contents

vi   **Contents**

Participants

Session I

Ursula K. Abbott, Chairperson
Department of Avian Sciences, University of California, Davis, CA 95616
W. Scott Argraves [15]
Department of Animal Genetics, University of Connecticut, Storrs, CT 06268
Arnold I. Caplan [37]
Developmental Biology Center, Biology Department, Case Western Reserve University, Cleveland, OH 44106
Paul F. Goetinck [15]
Department of Animal Genetics, University of Connecticut, Storrs, CT 06268
Peggy L. Lever-Fischer [15]
Department of Animal Genetics, University of Connecticut, Storrs, CT 06268
Paula J. McKeown-Longo [15]
Department of Animal Genetics, University of Connecticut, Storrs, CT 06268
Jane M. Oppenheimer [1]
Biological Laboratories, Bryn Mawr College, Bryn Mawr, PA 19010
Mitchell I. Quintner [15]
Department of Animal Genetics, University of Connecticut, Storrs, CT 06268
Linwood M. Sawyer [15]
Department of Animal Genetics, University of Connecticut, Storrs, CT 06268
Kenneth J. Sparks [15]
Department of Animal Genetics, University of Connecticut, Storrs, CT 06268

Session II

Arthur Chovnick, Chairperson
Department of Genetics, University of Connecticut, Storrs, CT 06268
N.V. Fedoroff
Department of Embryology, Carnegie Institution of Washington, Baltimore, MD
C. Weldon Jones [69]
Cellular and Developmental Biology, Biological Laboratories, Harvard University, Cambridge, MA 02138
Fotis C. Kafatos [69]
Cellular and Developmental Biology, Biological Laboratories, Harvard University, Cambridge, MA 02138
Gerald Rubin
Department of Tumor Virology, Sidney Farber Cancer Institute, Boston, MA

The number in brackets following a participant's name is the opening page number of that participant's paper.

vii

Session III

David C. Baulcombe [83]
Department of Botany, University of Georgia, Athens, GA 30602
Paul B. Green, Chairperson
Department of Biological Sciences, Stanford University, Stanford, CA 94305
Tom J. Guilfoyle [99]
Department of Botany, University of Minnesota, St. Paul, MN 55108
Joe L. Key [83]
Department of Botany, University of Georgia, Athens, GA 30602
Philip A. Kroner [83]
Department of Botany, University of Georgia, Athens, GA 30602
C.S. Levings III [119]
Department of Genetics, North Carolina State University, Raleigh, NC 27650
M.G. Neuffer [137]
Department of Agronomy, University of Missouri, Columbia, MO 65211
R.R. Sederoff [119]
Department of Genetics, North Carolina State University, Raleigh, NC 27650
William F. Sheridan [137]
Department of Biology, University of North Dakota, Grand Forks, ND 58202
Linda L. Zurfluh [99]
Department of Botany, University of Minnesota, St. Paul, MN 55108

Session IV

Ann E. Blechl [185]
Laboratory of Genetics, University of Wisconsin, Madison, WI 53706
Eugene T. Butler III
Division of Biology, California Institute of Technology, Pasadena, CA 91125
Lewis Cantley [171]
Department of Biochemistry and Molecular Biology, Harvard University, Cambridge, MA 02138
Ross C. Hardison
Division of Biology, California Institute of Technology, Pasadena, CA 91125
David Housman [171]
Center for Cancer Research and the Department of Biology, Massachusetts Institute of Technology, Cambridge, MA 02139
Vernon M. Ingram, Chairperson [157]
Department of Biology, Massachusetts Institute of Technology, Cambridge, MA 02139
Robert W. Keane [157]
Department of Biology, Massachusetts Institute of Technology, Cambridge, MA 02139
Elizabeth H. Lacy
Division of Biology, California Institute of Technology, Pasadena, CA 91125

Robert Levenson [171]
Center for Cancer Research and the Department of Biology, Massachusetts Institute of Technology, Cambridge, MA 02139
Tom Maniatis
Division of Biology, California Institute of Technology, Pasadena, CA 91125
Che-Kun James Shen
Division of Biology, California Institute of Technology, Pasadena, CA 91125
Sixiang Shen [185]
Laboratory of Genetics, University of Wisconsin, Madison, WI 53706
Jerry L. Slightom [185]
Laboratory of Genetics, University of Wisconsin, Madison, WI 53706
Oliver Smithies [185]
Laboratory of Genetics, University of Wisconsin, Madison, WI 53706
Elio F. Vanin [185]
Laboratory of Genetics, University of Wisconsin, Madison, WS 53706

Session V

Anne McLaren [201]
MRC Mammalian Development Unit, Wolfson House, 4 Stephenson Way, London NW1 2HE, England
Takeshi Matsunaga [235]
Division of Biology, City of Hope Research Institute, Duarte, CA 91010
Susumu Ohno [235]
Division of Biology, City of Hope Research Institute, Duarte, CA 91010
M.H.L. Snow [201]
MRC Mammalian Development Unit, Wolfson House, 4 Stephenson Way, London NW1 2HE, England
Stephen Subtelny, Chairperson
Department of Biology, Rice University, Houston, TX 77005
P.P.L. Tam [201]
MRC Mammalian Development Unit, Wolfson House, 4 Stephenson Way, London NW1 2HE, England
Stephen S. Wachtel [219]
Division of Cell Surface Immunogenetics, Memorial Sloan-Kettering Cancer Center and Division of Pediatric Endocrinology, New York Hospital-Cornell Medical Center, New York, NY 10021

Participants at the Thirty-Ninth Symposium of the Society for Developmental Biology

Participants at the Thirty-Ninth Annual Symposium of the Society for Developmental Biology

1. William Sheridan
2. Bert Gold
3. Raymond E. Sicard
4. Mary F. Lombard
5. Ursula K. Abbott
6. Fritz Schwalm
7. Arnold J. Kahn
8. Hannah Friedman Elson
9. Carolyn Conway
10. John McCarrey
11. Lynn Lamoreux
12. Frank Tufaro
13. David Gwynne
14. Mike Snow
15. Marie Dziadek
16. Anna Seitz
17. Mary Kujawa
18. M.G. Neuffer
19. William Laffond
20. Gerald Kidder
21. Marcelo Jacobs-Lorena
22. Stephen Subtelny
23. Eileen Hickey
24. Elliott Goldstein
25. Jane Oppenheimer
26. Not identified
27. Robert W. Keane
28. John G. Scandalios
29. Michael Bradbury
30. Not identified
31. Deborah L.G. Brodie
32. Robert A. Burns
33. Dorothea Rudnick
34. George P. Livi
35. D.F. Poulson
36. Mary Clutter
37. Susan Ernst
38. Laura Gillespie
39. Cynthia Fisher
40. Not identified
41. Barbara Johnson-Wint
42. Ilse Schwinck
43. Frances T. Hakim
44. Raziel S. Hakim
45. Not identified
46. Not identified
47. Jim Morrissey
48. Paul Goetinck
49. Tom Guilfoyle
50. Phillip Osdoby
51. Meredith Runner
52. Cheryl Anderson
53. Robert Kelley
54. Ben Murray
55. Holly Schauer
56. Robert L. Searls
57. Mike O'Guin
58. Marie DiBerardino
59. John Fallon
60. Carl McDaniel
61. Linda Hyman
62. Bob McGaughey
63. Thomas Barnett
64. George Dearlove
65. Gary W. Grimes
66. Vijay Thadani
67. Steve Barclay
68. Paula Mayerson
69. Ken Fisher
70. Alice L. Bull
71. Not identified
72. Mark Goldie
73. Joel Schindler
74. Not identified
75. Eleanor Wenger
76. Art Conway
77. Byron S. Wenger
78. Yiu-Kay Lai
79. Alice Wertheimer
80. David Waddell
81. A. Scott Goustin
82. Robert Trelsted
83. R.O. Erickson
84. Bill Loomis
85. Paul B. Green
86. James Pustell
87. Not identified
88. Nancy Marcus
89. Stephen Cohen
90. Leon Browder
91. Doug Easton
92. D. Baulcombe

Dedicated to the Memory of
Dr. Walter Landauer
Professor Emeritus, Animal Genetics
University of Connecticut

Young Investigator Awards — 1980

First Place Award

Xanthe Vafopoulou-Mandalos
Graduate Student
Biological Sciences Group
University of Connecticut
Hans Laufer, Sponsor

Juvenile Hormone and Ecdysterone Regulate Hemoglobin
Synthesis in Chironomus thumi

Second Place Awards

James H. Morrissey
Graduate Student
Biology Department
University of California, San Diego
William F. Loomis, Sponsor

Mutational Analysis of Pattern Formation in Dictyostelium discoideum

Vijay M. Thadani
Graduate Student
Department of Biology
Yale University
Clement L. Markert, Sponsor

A Study of Ooctye Maturation and Heterospecific Sperm-Egg Interaction in the
Rat, Mouse, and Deer Mouse Using In Vitro Fertilization and Sperm Injection

Abstract of the First Place
Young Investigator Award

Juvenile Hormone and Ecdysterone Regulate Hemoglobin Synthesis in Chironomus thummi (Diptera). Xanthe Vafopoulou-Mandalos and Hans Laufer, Biological Sciences Group, University of Connecticut, Storrs, CT 06268.

The hemolymph of Chironomus 4th instar contains as much as 80% soluble polymorphic hemoglobins (Hbs). The Hb content of larval hemolymph decreases to 46% during molting from the 3rd to the 4th instar and from the prepupa to pupa, to finally degrade in pupae. Cytochemical and biochemical studies indicate that Hb synthesis occurs primarily in the larval subepidermal fat body. Nine Hbs are resolved by 12.7% acrylamide gel electrophoresis. Hb synthesis in the presence ^{3}H-δ-aminolevulinic acid and ^{14}C-amino acids in vitro shows that Hbs 2 and 3 are stage specific for the 4th instar and are first detected by day-4 of this stage. In vivo and in vitro treatment with the potent juvenile hormone (JH) mimic Altosid induced early synthesis of Hbs 2 and 3 at day-2 of the 4th instar. Actinomycin D inhibits this early induction in vitro, while it does not affect Hb synthesis of mid-4th instar fat body. Ecdysterone, on the other hand, inhibits Hb synthesis of mid-4th instar tissue in the presence or absence of actinomycin D. The results are interpreted as follows: Normal Chironomus larvae possess long-lived Hb-messages. Molting from the 3rd to the 4th instar, regulated by ecdysterone, results in temporary depletion of Hb mRNAs. Reinitiation of Hb synthesis depends on new Hb mRNA synthesis induced by JH in the absence of ecdysterone. In pupation, Hb synthesis terminates because of ecdysterone secretion in the absence of JH.

(Supported in part by grants from the National Science Foundation and the University of Connecticut Research Foundation.)

Walter Landauer and Developmental Genetics

Jane M. Oppenheimer

Biological Laboratories, Bryn Mawr College, Bryn Mawr, Pennsylvania 19010

I. INTRODUCTION

Walter Landauer's name first appeared in the records of this Society, in his own hand, in June 1941, on the list of attendants at the Growth Symposium held that year in Hanover, New Hampshire; L. C. Dunn referred to some of Landauer's work in a paper he presented there [Dunn, 1941]. Meetings of what was then called the Society for the Study of Development and Growth were held in Storrs, Connecticut, in August 1947 (the Seventh Symposium) and again in June 1963 (the Twenty-second Symposium). In 1948, when the Eighth Symposium was held in Burlington, Vermont, Landauer delivered a talk on "Hereditary Abnormalities and their Chemically-induced Phenocopies" [Landauer, 1948b].

Levels of Genetic Control in Development, pages 1–13

Walter Landauer
July 15, 1896—February 18, 1978

The books of the Society in 1946 showed a charge of fifteen dollars for a gift for Landauer (otherwise unexplained: perhaps something to be set aside to be presented as a token of gratitude to him after the 1947 Storrs meeting?). He received other gifts, too: the Borden Award of a thousand dollars and a gold medal were given to him in 1954 by the Poultry Association for, among other things, delving "into the disciplines of genetics, physiology, nutrition and environment, exploring each as it affects the normal development of the chicken embryo" [Anon, 1954, p 1288]. We offer this meeting to his memory as a tribute to him for all he did to bring genetics and embryology into close relationships with each other and with biochemistry, and to illuminate genetics itself.

The final paragraph of his 1948 talk to this Society concluded as follows:

"It should be possible with chemical phenocopy methods to trace the initiation of gene effects to earlier stages than is possible visually and to localize these effects more accurately. It should also be possible to study means of counteracting the chemical induction of phenocopies. Positive results in this respect would open the way for a direct comparison of the biochemical effects of mutant genes and phenocopies" [Landauer, 1948b, p 196]. This was not Landauer's first emphasis on phenocopies; Goldschmidt had introduced the term and concept in 1935, and Landauer expressed its importance for his own work at least as early as 1946 when, in connection with the induction of rumplessness by insulin, he noted that his phenocopies showed a number of characteristics similar to genetic traits [Landauer and Bliss, 1946]. By 1952 he expressed as his ultimate aim the testing of the substances effecting protection against phenocopy production on the copied mutant stocks to ascertain whether a similar beneficial result might be obtained [Landauer, 1952a]. The contents of the present paper cover some of the backgrounds for, and some of the results achieved by, Landauer's work. It was so varied and full, and his pen so prolific, that it is not possible even to touch upon his total output in the space allotted.

II. SOME BIOGRAPHICAL COMMENTS

A. Skeletal Details

Landauer was born in Mannheim, Germany, on July 15, 1896. He was a Red Cross nurse during World War I. Then he studied, as did so many young Germans of his time, at more than one university; he went first to Frankfurt am Main, then to Heidelberg, where in 1922 he took his degree (Dr Phil Nat) with Curt Herbst, whom he admired very much. The subject of his dissertation was the influence of ammonia on the shift of direction of heredity in hybrid sea-urchin larvae [Landauer, 1922]. After two years as an assistant in the Zoological Institute in Heidelberg, he emigrated to the United States in 1924; he was naturalized in 1940. Storrs was his first destination here. He was Assistant Geneticist at the Storrs Agricultural Experiment Station from 1924 to 1928, then Professor of Genetics at the University of Connecticut in Storrs from 1928 to 1964. Upon his retirement in 1964 he removed to London, where he held an honorary appointment first in the Department of Zoology and Comparative Anatomy in University College, London, then from 1967 until his death, on February 18, 1978, in the Department of Animal Genetics. He married Elly T. Bernstein in London on September 4, 1964, and it is to her that we transmit this Memorial with admiration and respect.

B. Biographical Bones Fleshed Out

How did it come about that a young man moved within two years from sea-urchin hybridization to poultry breeding? The answer is not completely apolitical. Disorders in Germany during the Weimar Republic, which extended even so far—or so near—as the murder of a much-loved Uncle, Gustav Landauer, gave Walter early insight into the future of Germany. Accordingly, when Leslie C. Dunn, already in Storrs for several years, wrote to Richard Goldschmidt for suggestions as to a possible German-trained post-doctoral assistant, he soon received a cable saying "Determined to come. Walter Landauer" (from letter of March 30, 1980 Mrs. Leslie C. Dunn to Jane Oppenheimer). Mrs. Dunn went on to say that there then followed some direct correspondence between the principals, and in due time Walter appeared "at the railroad station in Willimantic, his luggage consisting mostly of a microscope and a calf's head in formaldehyde, leaking a bit" (ibid). Dunn and Landauer soon embarked on a study of lethal genes in poultry that was to be continued by Landauer until the end of his life (his last paper was published posthumously).

III. POULTRY BREEDING

Artificial selection of fancy breeds of fowl for exhibition was an ancient practice by the time Mendel's laws were rediscovered. Darwin devoted a chapter to it in *The Variation of Animals and Plants under Domestication* [Darwin, 1868, 1892 ed]: he included as numbers 8, 9, 10, and 11 in his list of fancy breeds "Bantams (originally from Japan), . . . Rumpless Fowls, . . . Creepers or Jumpers, . . . and Frizzled or Caffre Fowls" [Vol I, p 211]; these were all to be of interest to Landauer. I believe it was Charles Davenport [1906] who first wrote on inheritance in poultry in mendelian terms in this country, including mention of a ɔross involving Frizzle fowls. In fact, Frizzle fowls had been known at least as far back as 1600, when Aldrovandus [1600, 1963 ed] published a picture of one: Landauer knew about Aldrovandus' interest in them.

The first lethal factor was discovered in 1905, not in fowl but in the yellow mouse [Cuénot, 1905] and was first correctly interpreted by Castle and Little in 1910. Dunn, in 1923, just three years after coming to Storrs, reported the first lethal gene found in poultry, a recessive in White Wyandottes that was subsequently lost. So Landauer appeared at an interesting moment. At the meetings of the American Society of Zoologists in 1924, I. E. Cutler read a paper on "Reptilian Fowls" with leg and wing bones shorter and thicker than those of normal chickens [Cutler, 1925]; it was he who provided Dunn and Landauer with birds to establish their own flock; Dunn and Landauer began writing about the genetics of Creepers in 1926 [Dunn and Landauer, 1926]. By 1930 they had also established flocks of German, Scottish, and Marquesan origin [Landauer and Dunn, 1930b], and by 1940 Japanese bantams, too [Landauer, 1942]. Landauer in due time would study Creeper-like characters in Ancon sheep, African dwarf goats, Dexter cattle, and certain strains of ducks and mice; he also kept dachshunds for house pets. The Ark was his supply house.

While their Creeper flocks were becoming established, Landauer and Dunn were studying other mutations. Dunn [1925] described the inheritance of a mutation for rumplessness in fowl, and immediately thereafter published jointly with Landauer an article comparing hereditary with so-called accidental rumplessness [Landauer and Dunn, 1925]. This mutation for rumplessness was dominant; in due time an apparently recessive gene for the same trait was also discovered [Landauer, 1945b]. Rumplessness was later to be one of the principal traits studied by Landauer in phenocopies.

A third character of particular interest to Dunn and Landauer was the Frizzle trait; they started to breed Frizzle fowl in 1927. These birds have curly body feathers, and the homozygotes are often bare for their early years of life, and what feathers they do develop are scarce and scraggly. Landauer and Dunn [1930a] established that the responsible gene is a single dominant, and they gave detailed microscopical descriptions of the defects of the feathers.

These were not the only mutations that Landauer studied, to the benefit of developmental genetics, but it is not possible even to list, let alone to discuss them all; investigations of the three I have selected will be illustrative of how he carried out some of his more important work. Dunn left Storrs in 1928, though he remained the Station's Consulting Geneticist, and joint papers by Landauer and Dunn continued to appear at least until 1936. Landauer's later collaborators in related fields were to continue to be of high distinction: Honor Fell, when whole bone rudiments were cultured; F. G. Benedict, when basal metabolism was studied; T. C. Byerly, when nutritional deficiencies brought about developmental abnormalities; B. Zondek and C. A. Pfeiffer and W. U. Gardner, when estrogens were used; S. D. Aberle, when other endocrine glands were considered; R. A. Fisher, when there was a problem of crossing-over in close linkage; C. I. Bliss, when statistical problems were at issue. These are only some examples of Landauer's stellar collaborators, and his acknowledged advisors were of equal distinction. None but the best would do.

This section might well be concluded by brief mention about the quality of his own experimental methodology. His breeding experiments were impeccable; he knew how to maintain proper controls; he used appropriate statistical methods to analyze his data. While his early interpretations were questioned by his colleagues on occasion, or even by himself, they often proved self-correcting later [see especially Landauer, 1972].

IV. FRIZZLE FOWL AND CREEPERS

We will return to rumplessness when we discuss the phenocopy problem; meantime let us take up the contributions of Landauer's studies on Frizzle fowl and Creepers to an understanding of normal avian physiology, and of lethal actions of genes, respectively.

A. Frizzle Fowl

The birds carrying the Frizzle character sometimes reach reproductive age, thus this mutation is not an obligate lethal. Perhaps the most important contribution of the Frizzle fowl is the fact that, by its very existence, it elucidates factors controlling metabolism under altered temperature relations; how convenient for physiologists to find a chicken with a defective feather covering! Landauer and collaborators made an extensive study of the mutants' metabolic rate [Benedict et al, 1932]. They studied their blood cells and hemoglobin [Landauer and David, 1933], the effect of elevated metabolism on the heart [Boas and Landauer, 1933, 1934]; Landauer was particularly interested in the endocrine glands of Frizzle fowl [Landauer and David, 1934; Landauer and Aberle, 1935; Aberle and Landauer, 1935]. Had he never gone beyond these studies on Frizzle fowl he would have become well known as a physiologist. In fact, he did write the chapter on genetic aspects of physiology for the 15th edition of Howell's great textbook of medical physiology [Landauer, 1946].

B. Creepers

Perhaps the most excitement, in its time, derived from Landauer's studies on the effects of the Creeper gene on development, and it may well be that these articles attracted many bright young people into developmental genetics before it had become an organized science. Development and genetics were separate disciplines in the 1920s and 1930s, and the studies of the Creeper genes pulled them together in a most fascinating fashion, just as the studies on phenocopies were later to add biochemistry to the combination.

The Creeper gene is a dominant lethal [Cp, Dunn and Landauer, 1926; Landauer and Dunn, 1930b]. The birds that we see and call Creepers are heterozygotes (Cp/ +); their deficit is comparable to chondrodystrophy in man and other mammals [Dunn, 1927; Landauer, 1927]; in fact, in 1927 Dunn [ibid] was calling the heterozygotes chondros. Micromelia is the technical term for shortened limbs, and in 1969 Landauer published a bibliography of 1755 items on micromelia, each briefly annotated by himself [Landauer, 1969].

The Cp/Cp homozygotes are visibly retarded by between 36 and 72 days of incubation in ovo, and most of the embryos die during the 3rd or 4th days of incubation. A small percentage of homozygotes (around 6.5%), so-called escapers, live beyond this stage. These are phocomelic and also have colobomatous and otherwise anomalous eyes and other head parts. The percentage of escapers was raised to 17.2% by incubating the eggs at 35.6°C during the first 24 hours of development, instead of at the usual 38°–39°C [Landauer, 1944]. Fragments raised in vitro were able to survive beyond the usual lethal period [David, 1936]. Hamburger [1941] transplanted Cp/Cp limb rudiments at two to three days of incubation to normal hosts and found them to reach the phocomelic stage and age of the escapers' limbs; a few years later areas that were genetically destined to become Cp/Cp limbs were grafted at 24 to 35 hours of incubation, before the rudiments were visible and before the establishment of the circulation; the results were the same [Rudnick, 1945]. In contrast, when the Cp/Cp eye rudiments were grafted orthotopically to normal hosts, they developed normal eyes [Gayer, 1942; Gayer and Hamburger, 1943]. Also, when the American Creepers were mated with the Scotch, German, and Marquesan strains already mentioned [Landauer and Dunn, 1930b] and with Japanese bantams [Landauer, 1942], the numerical results were different. How some of these variations were explained is discussed later.

C. Some Interpretations

An early idea of Landauer [1932a] was that the homozygous Cp/Cp effect was due to nonspecific growth inhibition, and for a while this seemed confirmed by work in in vitro with Fell [Fell and Landauer, 1935] and by David [1936]. Early in the development of the work, Landauer thought the growth inhibition probably related to what he called a section deficiency in a chromosome [Landauer, 1933]. This was mentioned occasionally later but never followed up cytologically; it was supported by some crossover studies.

Transition was made from growth to biochemical agents as causal factors very gradually. "Especially, if as is likely, the Creeper mutation should prove to be a deficiency, it seems probable that the genetic difference is responsible for the failure of some substance, necessary for growth, being available at the right time," Landauer wrote in 1934 [1934a, p 32]. But the emphasis remained on growth. Had the growth substance concept been incubating in his mind since his early years with Herbst?

By 1939, when he addressed the Seventh International Genetics Congress in Edinburgh, his tune began to vary, and the interpretation that the fate of parts is determined by alterations in the embryo as a whole began to be supplanted by the idea that "the localized effects are due merely to quantitative differences in physiological needs of certain parts at definite periods; these differences depending on such factors as differential growth activity, involving different needs for certain enzymes and other substances" [1941, p 183].

V. PHENOCOPIES

Long before Landauer had begun to take up phenocopies as such—in fact, before they were even defined—he was studying the effect of artificially introduced teratogens. In 1929 he reported on work, begun in 1925, on the toxic action of lithium and magnesium on chick development [Landauer, 1929]; his early mentor, Herbst, had found lithium salts to disturb sea urchin development in 1892 [Herbst, 1892]. In 1932 Landauer irradiated chicken eggs with ultraviolet light [Landauer, 1932b]; in 1934 he studied the effects of vitamin D deficiency on chicken development [Landauer, 1934b]. He studied chick micromelia brought about by nutritional deficiency [Byerly et al, 1935]; he shook eggs prior to incubation [Landauer and Baumann, 1943]. He never forgot the influence of the environment on development or on genes.

A. Insulin as a Teratogen

"Incidental to other investigations it was observed that the injection of solutions of certain chemicals led to the appearance of increased numbers of rumpless embryos and chicks. It was decided to make a systematic study of this problem;" thus began Landauer's first paper on rumplessness produced by the injection of insulin and other chemicals [1945a, p 65]. The latter included at first cysteine hydrochloride, cystine, dl-glutamic acid hydrochloride, thioglycollic acid, and l-malic acid; d-malic acid was without effect. Inactivation of the insulin abolished the rumplessness-inducing effect; reactivation restored it [Landauer and Lang, 1946]. When insulin was used, if the dosages, the times of administration, and the stocks used for the experiments (Creeper or White Leghorn) were varied, the results differed quantitatively. It was concluded that "the genotype of particular stocks may modify the 'penetrance' as well as the 'expressivity' of insulin-induced rumplessness. In this respect our phenocopies show a behavior similar to that of many genetic traits" [Landauer and Bliss, 1946, p 21]. By this time it seemed likely to Landauer [1947b, p 325] that the insulin acts on one major physiological chain of events. But it was soon demonstrated that when insulin was injected later in development, the results differed qualitatively also; micromelia resulted [Landauer, 1947a]. Eventually insulin was shown also to produce antiteratogenic effects [Landauer, 1972].

By 1948—the same year that Landauer spoke so eloquently at the Growth Symposium in Burlington—he was attempting to prevent the teratogenic effects of insulin by the use of nicotinamide and ketoglutamic acid; the nicotinamide proved effective, ketoglutamic acid less so. Zwilling [1948] showed that insulin produced hypoglycemia in early embryos in degree and duration parallel to the extent of micromelia, and Landauer believed then that the nicotinamide effect provided strong evidence that the insulin-induced abnormalities are mediated by disturbances of the embryos' carbohydrate metabolism [Landauer, 1948a]. This seemed further confirmed when pyruvic acid, if administered during anaerobic stages of development, prevented insulin damage; the interpretation in 1952 was that the teratogenic effects of insulin are produced by rendering coenzyme unavailable to the embryo. "Supplements to insulin treatment seem to be beneficial or noxious according to whether they aid or hinder in relieving the insulin determined scarcity of coenzyme" [Landauer and Rhodes, 1952, p 259].

B. Other Teratogens

Landauer had earlier experimented with riboflavin as a supplement for insulin treatment, with negative results; in order to obtain a more concentrated solution, boric acid was used as a solubilizer of riboflavin, and lo, boric acid itself proved teratogenic, producing rumplessness if injected at early stages (though less frequently than insulin), other anomalies when administered later. The skeletal malformations were similar to those known to result from riboflavin deficiency [Landauer, 1952b]; supplementation of early boric acid treatment with riboflavin significantly lowered the incidence of rumplessness [ibid].

Among what seems an almost endless number of compounds tested, pilocarpine hydrochloride was the next to be singled out for more intensive study [Landauer, 1953]. Again, when administered in early stages, it produced rumplessness, among other anomalies; nicotinamide employed as a supplement offered protection. Again, carbohydrate utilization seemed to be interfered with [Landauer, 1954].

Although the effects of the substances used to produce the phenocopies differed in a number of respects in details of expression [Landauer, 1954], the use of those named here and of many other compounds over the years led to the conclusion that interference with oxidative phosphorylation is the cause of many teratogenic effects [Landauer and Wakasugi, 1967]. Interference with purine synthesis or with functions of NAD-linked dehydrogenases were pinpointed as specific effective causes [Landauer and Wakasugi, 1968]; Landauer and Sopher [1970] emphasized what they considered clear evidence for the importance of interference with mitochondrial energy in the production of malformations. This series of works seemed to conclude when, in one of his most brilliant papers, Landauer [1972] presented an exhaustive critique of the literature on the formerly supposed relation between hypoglycemia and teratogenicity in the origin of malformations resulting from treatment with insulin and certain sulfonamides.

A new and final set of experiments, using cholinomimetic drugs, was begun in 1975 [Landauer, 1975a]. Their aim was to study muscular hypoplasia, one effect of the administration of insulin plus 3-acetyl pyridine in 1972. Landauer now wished to investigate why distinct teratogens produce deficits that seem unrelated morphologically and metabolically. When the cholinominimetic substances were used as teratogens, it seemed that neuromuscular events played a decisive role in the results [Landauer, 1975a, b, 1976, 1977, 1978; Landauer et al, 1976]. Landauer's ultimate purpose, originally articulated in 1952, of testing the action in the original mutants of substances demonstrated as protective in their phenocopies, was never fulfilled in his lifetime. Now direct gene-tampering seems more appealing as a therapeutic venture.

VI.   NORMAL GENE ACTION

A. Modifiers

Among Landauer's outstanding traits were his humanity and kindness, and he can never have forgotten that studies in teratology might have benefits for unfortunate creatures, including men, suffering from genetic anomalies. But he also never forgot that he was utilizing his phenocopies to study normal genetics. His interest in Frizzle fowl as instructors in physiology provides only one example. The effects of his teratogenic and antiteratogenetic agents were often complicated, just as were those of the mutant genes he studied: remember the Creeper escapers! Penetrance and expressivity were factors that he kept in mind in all of his analyses. They were affected by dosages, stages, times, and duration of treatments when foreign substances were used, by general and special environmental factors, and by the genetic backgrounds of the stocks he used for any experiment or group of experiments.

As a result he repeatedly put strong emphasis on the action of modifying genes. "The existence of numerous modifiers or suppressors, which in the presence of mutant genotypes, canalize development toward 'normal', has been a common experience in our studies with fowl," he wrote in 1965 [Landauer, 1965, p 136]. The plasticity of the genome, striving always to regulate towards the normal, is one of the constant themes that runs through much of his work. When lecturing "On Phenocopies, Their Developmental Physiology and Genetic Meaning," he concluded that "taking all evidence into account, it seems to me that the most exciting aspect of the study of phenocopies is the opportunity it may provide of shedding light, if only indirectly, on the developmental functions of that awesome skeleton in the closets of genetical science—the normal genotype" [Landauer, 1958, p 211].

B. Hatchability

Insofar as direct utility is concerned, one of Landauer's most important contributions may have been assembling data on chick hatchability. In 1937 a small monograph appeared on "The Hatchability of Chicken Eggs as Influenced by Environment and Heredity" as Bulletin 216 of the Storrs Agricultural Experiment Station; it had 84 pages. New revisions appeared in 1941 (124 pages), in 1948 (231 pages), in 1951 (223 pages), in 1961 (278 pages), and in 1967 (315 pages). The citation for the Borden Award claimed in 1954 that "no one investigator has contributed more to an understanding of the factors influencing the hatchability of chicken eggs. . . . This bulletin will long remain an authoritative text" [Anon, 1954, p 1288], and so it has done.

VII.   ENVOI

Landauer's tirelessness and energy permitted him to carry out many other activities in and out of the laboratory and chicken coop; he was a prodigious reader; he was interested in history; he was interested in music; he was interested in art. His collection of Kaethe Kollwitz' works numbered over 100 prints and drawings, left to the William Benton Museum of Art of the University of Connecticut. Between 1919 and 1929 he wrote over 60 articles of political intent that appeared in the public press (list of references provided by Dorothea Rudnick).

His outstanding gifts were, however, not only his energy and his intelligence, but also his utter humaneness. When Landauer wrote Punnett's obituary, he concluded it by saying: "He was a brilliant, a good and a kind man who faced the exigencies of life with courage and equanimity" [1967, p 766]. That is what we say of Landauer.

ACKNOWLEDGMENTS

I am grateful for information and assistance to Ursula Abbott, Ruth Bellairs, Edward Buss, Louise Dunn, Viktor Hamburger, Holly Schauer, and especially to Paul Goetinck and Dorothea Rudnick.

REFERENCES

Aberle SD, Landauer, W (1935). Anat Rec 62:331–335.
Aldrovandus O (1600; 1963 ed). "Aldrovandi on Chickens." (CR Lind, trans and ed), University of Oklahoma Press, Norman, Oklahoma.
Anon (1954). Poultry Sci 33:1288.
Benedict FG, Landauer W, Fox EL (1932). Bull #177 Storrs Agr Exp Sta, 101 pp.
Boas EP, Landauer W (1933). Am J Med Sci 185:654–665.
Boas EP, Landauer W (1934). Am J Med Sci 188:359–364.
Byerly TC, Titus HW, Ellis NR, Landauer, W (1935). Proc Soc Exp Biol Med 32:1542–1546.
Castle W, Little CC (1910). Science 32:868–870.
Cuénot L (1905). Arch Zool Exp Gén Sér 4 3:123–132.
Cutler IE (1925). J Hered 16:352–356.
Darwin CR (1868; 1892 ed). "The Variation of Animals and Plants under Domestication." 2 vols. D Appleton, New York.
Davenport C (1906). "Inheritance in Poultry." Washington, DC: Carnegie Institution of Washington.
David PR (1936). Wilhelm Roux Arch 135:521–551.
Dunn LC (1923). Am Nat 57:345–349.
Dunn LC (1925). J Hered 16:127–134.
Dunn LC (1927). Wilhelm Roux Arch 110:341–365.
Dunn LC (1941). In "Third Growth Symposium." Growth 5(Suppl):147–161.
Dunn LC, Landauer W (1926). Am Nat 60:574–575.
Fell HB, Landauer W (1935). Proc R Soc Lond Ser B 118:133–154.
Gayer K (1942). J Exp Zool 89:103–133.
Gayer K, Hamburger V (1943). J Exp Zool 93:147–180.
Goldschmidt R (1935). Z Vererbungslehre 69:38–69; 70–131.
Hamburger V (1941). Physiol Zool 14:355–365.
Herbst C (1892). Z wiss Zool 55:446–518.
Landauer W (1922). Wilhelm Roux Arch 52:1–94.
Landauer W (1927). Wilhelm Roux Arch 110:195–278.
Landauer W (1929). Poultry Sci 8:301–312.
Landauer W (1932a). J Genet 25:367–394.
Landauer W (1932b). Bull #179 Storrs Agr Exp Sta, 23 pp.
Landauer W (1933). Z mikrosc-anat Forsch 32:359–412.
Landauer W (1934a). Bull #193 Storrs Agr Exp Sta, 79 pp.
Landauer W (1934b). Am J Anat 55:229–252.
Landauer W (1941). Proc 7th Int Congress Genet Edinburgh, Scotland, pp 181–185.

Landauer W (1942). Am Nat 76:1–10.
Landauer W (1944). Science 100:1–3.
Landauer W (1945a). J Exp Zool 98:65–77.
Landauer W (1945b). Genetics 30:403–428.
Landauer W (1946). In "Howell's Textbook of Physiology," Ed 15. (JF Fulton, ed.) Ch 56. W. B. Saunders, Philadelphia.
Landauer W (1947a). J Exp Zool 105:145–172.
Landauer W (1947b). J Exp Zool 105:317–328.
Landauer W (1948a). J Exp Zool 109:283–290.
Landauer W (1948b). In "Eighth Growth Symposium." Growth 12(Suppl):171–200.
Landauer W (1952a). Ann NY Acad Sci 55:172–176.
Landauer W (1952b). J Exp Zool 120:469–508.
Landauer W (1953). J Exp Zool 122:469–483.
Landauer W (1954). J Cell Comp Physiol 43(Suppl 1):261–305.
Landauer W (1958). Am Nat 92:201–213.
Landauer W (1965). J Hered 56:131–138.
Landauer W (1967). Nature 213:766.
Landauer W (1969). In "Limb Development and Deformity: Problems of Evaluation and Rehabilitation." (CA Swinyard, ed.), pp 540–621. Charles Thomas, Springfield, Illinois.
Landauer W (1972). Teratology 5:129–136.
Landauer W (1975a). Teratology 12:125–145.
Landauer W (1975b). Teratology 12:271–276.
Landauer W (1976). Teratology 13:41–46.
Landauer W (1977). Teratology 15:33–42.
Landauer W (1978). Teratology 17:335–339.
Landauer W, Aberle SD (1935). Am J Anat 57:99–134.
Landauer W, Baumann L (1943). J Exp Zool 93:51–74.
Landauer W, Bliss CI (1946). J Exp Zool 102:1–22.
Landauer W, David LT (1933). Folia Haematol 50:1–14.
Landauer W, David LT (1934). Arch Int Pharmacodyn Thér 49:125–129; 130–143.
Landauer W, Dunn LC (1925). J Hered 16:153–160.
Landauer W, Dunn LC (1930a). J Hered 21:290–305.
Landauer W, Dunn LC (1930b). J Genet 23:397–413.
Landauer W, Lang EH (1946). J Exp Zool 101:41–50.
Landauer W, Rhodes MB (1952). J Exp Zool 119:221–261.
Landauer W, Sopher D (1970). J Embryol Exp Morphol 24:187–202.
Landauer W, Wakasugi N (1967). J Exp Zool 164:499–516.
Landauer W, Wakasugi N (1968). J Embryol Exp Morphol 20:261–284.
Landauer W, Clark EM, Larner MV (1976). Teratology 14:281–285.
Rudnick D (1945). J Exp Zool 100:1–17.
Zwilling, E (1948). J Exp Zool 109:197–214.

Chondrogenesis in Normal and Mutant Avian Embryos

Paul F. Goetinck, Peggy L. Lever-Fischer, Paula J. McKeown-Longo, Mitchell I. Quintner, Linwood M. Sawyer, Kenneth J. Sparks, and W. Scott Argraves
Department of Animal Genetics, The University of Connecticut, Storrs, Connecticut 06268

Levels of Genetic Control in Development, pages 15–35
© 1981 Alan R. Liss, Inc., 150 Fifth Avenue, New York, NY 10011

I. INTRODUCTION

In 1952, Dr. Landauer presented to the New York Academy of Sciences a paper titled *The Genetic Control of Normal Development in the Chicken Embryo* [Landauer, 1952]. That title encompasses his entire research interest. Dr. Landauer was particularly interested in the avian limb as a developmental system, and this interest was evident from his studies on hereditary limb defects and also from his studies on phenocopies, those non-hereditary abnormalities that resemble mutant traits. As part of both aspects of those studies he actively searched for mutants and adopted those whose propagation might not have been continued in other laboratories. He maintained these mutant flocks for future investigations on the genetic control of development because he strongly believed that "in order to be of real promise, such investigations (on the genetic control of development) must combine the methods of the geneticist, the embryologist, and the biochemist." We have combined these three approaches in our studies on cartilage differentiation.

Chondrocytes can differentiate from either mesenchymal or mesectodermal precursor cells. Given such cells with two different developmental histories, one can compare, by some parameter, the two precursor cell types, the two differ-

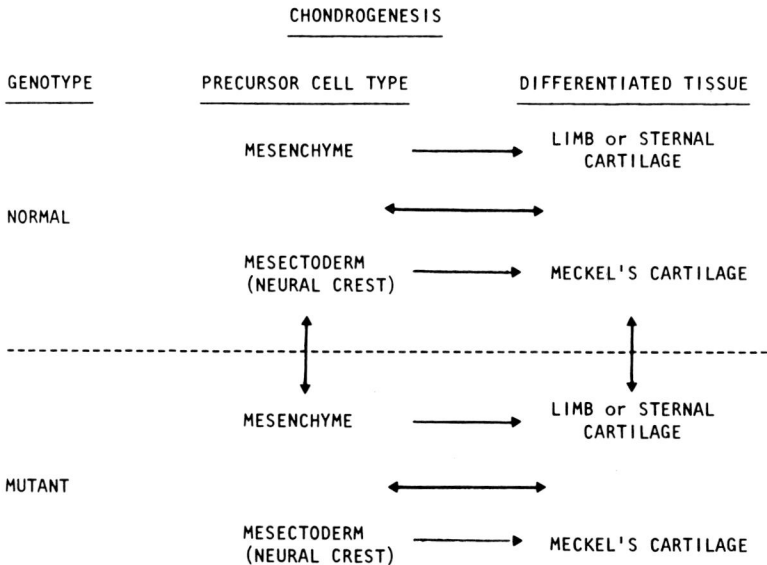

Fig. 1. Diagrammatic summary of the differentiation steps (single-headed arrows) and possible experimental comparisons (double-headed arrows) during chondrogenesis.

entiated tissues, and also the precursor cell types with their differentiated coun- terparts. Having, in addition, mutations that affect the development of cartila- ginous organs, one can further investigate chondrogenesis by making comparisons between the precursor cell types and the differentiated tissues of different genotypes. Figure 1 summarizes these differentiation steps (single- headed arrows) and the possible comparisons (double-headed arrows). The pres- ent report will review our studies on two mutants in which the development of cartilage is affected.

II. MUTANTS

A. Nanomelia (nm)

This mutant was discovered by Landauer [1965a] in descendants from crosses involving White Leghorns and Buff Cochin fowl. The mutation is inherited as a single recessive lethal autosomal gene. The nanomelic embryos are charac- terized by having shortened limbs and a parrot-like beak (Fig. 2). In view of the extreme reduction in the length of the limbs in these embryos, Landauer assigned to this mutation the name *nano*melia rather than *micro*melia, the usual term for shortened limbs.

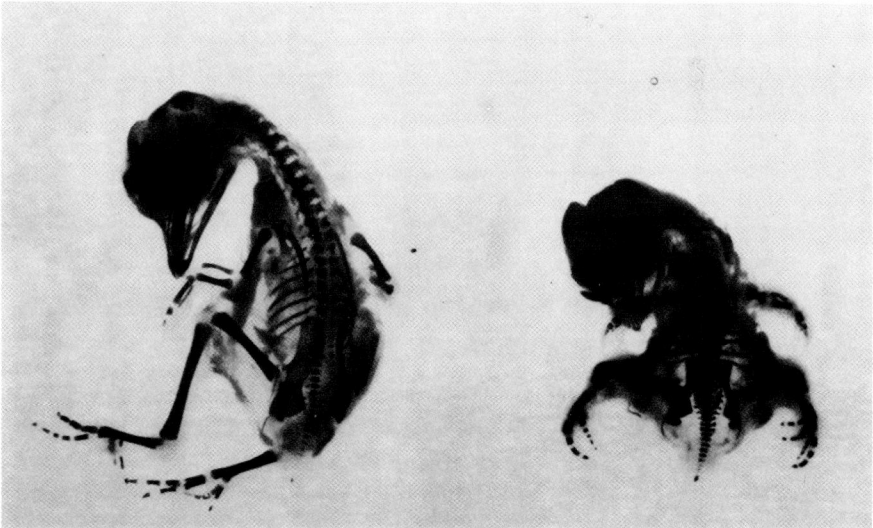

Fig. 2.   Alizerin red preparation of 17-day normal (left) and nanomelic (right) embryos.

B. Micromelia-Abbott (mm^A)

This mutation is also inherited as a single autosomal recessive gene, but it is genetically distinct from the nanomelia mutation. The mutation was discovered by U. K. Abbott at the University of California at Davis and was named by Landauer [1965b]. The most frequent phenotypic expression in embryos homozygous for this recessive lethal gene consists of a reduced size of the long bones of the limbs, a parrot-like beak, and a general overall reduction in body size. These embryos also have a hemorrhagic skin and a retarded rate of feather development. In addition to this most frequently expressed form of micromelia, there exists a form that has an intermediate phenotypic expression (mm^AI) (Fig. 3).

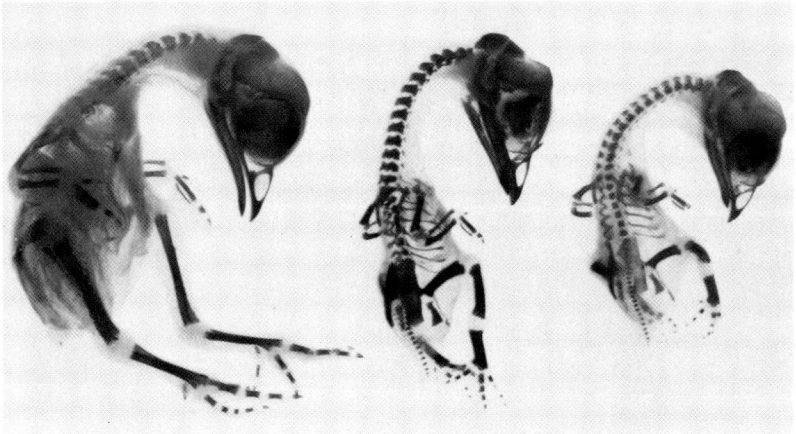

Fig. 3.   Alizerin red preparation of 17-day normal (left), micromelia-Abbott intermediate (middle), and micromelia-Abbott (right) embryos.

III. CARTILAGE DEVELOPMENT

A. The Limb

Limb buds of the avian embryo arise from the thickening of the somatopleure lateral to the somites at stage 16–17 of Hamburger and Hamilton [1951]. Up to stage 22, the limb mesoderm appears homogeneous by histological criteria. At about stage 22–23, primordia of the cartilaginous skeleton can be recognized as condensations in the core of the limb [Fell and Canti, 1934]. At stage 35, overt cartilage differentiation is evident.

B. Sternum

The sternum of birds first appears as a pair of mesodermal condensations, which later differentiate to form two separated cartilaginous plates. These sternal plates move toward the mid-line and fuse with each other to form the embryonic sternum [Fell, 1939].

C. Meckel's Cartilage

The chondrocytes in the limb and the sternum are of mesodermal origin. In contrast, chondrocytes of Meckel's cartilage and the hypobronchial skeleton are derived from the neural crest, which originates in the ectoderm [Johnston, 1966; Le Lièvre, 1974; Le Lièvre and Le Douarin, 1975]. Meckel's cartilage serves as the embryonic mandible, the proximal portion of which gives rise to the articular bone of the adult mandible.

IV. MACROMOLECULAR CHARACTERIZATION OF CARTILAGE DIFFERENTIATION

The synthesis of collagen and sulfated proteoglycans (PGS) has been studied extensively as a characteristic of the acquisition of specialized function associated with chondrocyte differentiation.

A. Collagen

The type of collagen that is synthesized by chondrocytes is distinct from that synthesized by other tissues. This initial finding made by Miller and Matukas [1969] and Trelstad et al [1970] has been confirmed and extended by a number of investigators [Miller, 1977]. All collagens contain three polypeptides. The collagen of skin and bone (type I) consists of two $[\alpha_1 \text{ (I)}]$ chains and one α_2 chain. All other collagens have three identical chains. Cartilage collagen (type II) contains three $[\alpha_1 \text{ (II)}]$ chains which are genetically distinct from those of type I. In the limb, precartilaginous mesenchyme synthesizes type I collagen. During chondrogenesis the core of the limb stops synthesizing type I and begins to synthesize type II collagens. When cartilage is later replaced by bone, type II collagen synthesis ceases and type I collagen synthesis is resumed [Linsenmeyer, 1974; Linsenmeyer and Smith, 1975; Linsenmeyer et al, 1973; von der Mark et al, 1976].

B. Sulfated Proteoglycans (PGS)

Normal embryonic sternal cartilage (SC) contains two populations of PGS. One [PGS(SC)-I] represents about 90% of the total proteoglycans, and the other

[PGS(SC)-II], the remaining 10%. PGS(SC)-II has been shown to contain several subpopulations [Kimata et al, 1974]. PGS(SC)-I is a cartilage-specific proteoglycan that exists in an aggregated state with hyaluronic acid and link proteins. The interaction of the proteoglycan monomer is ionic and is stabilized by the addition of one or more link proteins. The PGS monomer consists of a core protein to which two types of sulfated carbohydrate side chains are covalently linked. These side chains are chondroitin sulfate (ChS) and keratan sulfate (KS). Chondroitin sulfate, a repeating unit of sulfated N-acetylgalactosamine and glucuronic acid, is linked to the core protein via a neutral sugar linkage region (xylose-galactose-galactose), which is linked to core protein by an O-glycosidic linkage between xylose and hydroxyl groups of serine. The two galactose residues are followed by a glucuronic acid residue and the alternating repeating units of sulfated-N-acetylgalactosamine and glucuronic acid. Keratan sulfate, a repeating disaccharide unit of galactose and N-acetylglucosamine, is linked, in cartilage, to protein via O-glycosidic linkages between N-acetylgalactosamine and the hydroxyl groups of serine and threonine. The N-acetylglucosamine and, variably, the galactose residue are sulfated.

The core protein of PGS can be divided into three distinct regions. Chondroitin sulfate and keratan sulfate chains are clustered in two of these regions. One end of the core protein, which is free of carbohydrate side chains, is the hyaluronic acid binding region. Adjacent to this is the KS-rich region, followed by the ChS region [reviewed in Hascall, 1977].

In addition to ChS and KS side chains, two classes of oligosaccharides have been described on PGS from embryonic chick cartilage and Swarm rat chondrosarcoma [De Luca et al, 1980; Lohmander et al, 1980]. One class of oligosaccharide is linked to core proteins by the same linkage as KS—ie, an O-glycosidic bond between N-acetylgalactosamine and hydroxyl groups of serine and threonine. The second class is rich in mannose and appears to be linked to core protein by N-glycosylamine linkages to asparagine.

V. IMMUNOCHEMISTRY OF CARTILAGE PROTEOGLYCANS

Over the past several years a number of immunochemical studies have been carried out with cartilage PGS [reviewed in Dorfman et al, 1980]. As a rule the results of these studies have been conflicting and do not lend themselves to a uniform interpretation. This is due in part to the use of different PGS extraction and purification procedures. Another problem has been the general practice of treating PGS with hyaluronidase in order to demonstrate antigen–antibody interactions. The need for treatment of PGS with hyaluronidase, which removes the ChS side chains, is ascribed to the belief that the high negative charge of the ChS chains prevents antigen–antibody interactions [Ho et al, 1977]. However, since proteoglycans are measured by [³⁵S]sulfate incorporation or uronic acid, the enzymatic digestion of the ChS also removes the major means of identifying PGS.

The antibodies used in the studies to be described here were elicited in rabbits by immunization with highly purified, enzymatically undigested, juvenile chick cartilage PGS-I. The antiserum contains IgG, which binds embryonic [³⁵S]sulfate-labeled cartilage-specific PGS-I. The binding of antibody to the [³⁵S]sulfate-labeled PGS-I does not require prior enzymatic removal of the ChS side chains [Sparks et al, 1980]. The PGS-antibody interaction is demonstrated using the precipitation method of Farr [1958]. The antibody binds specifically to essentially all the PGS(SC)-I and not to the minor component, PGS(SC)-II (Fig. 4). Binding of PGS(SC)-I takes place either in its monomeric or aggregated state (Fig. 5). The antiserum does not contain antibodies to PGS from skin or four-day limb buds, nor to a number of sulfated glycosaminoglycans. The an-

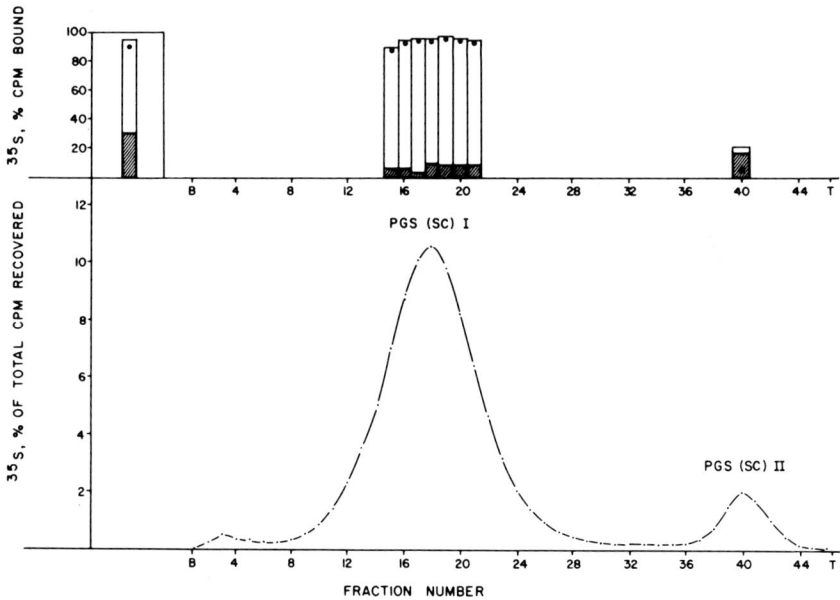

Fig. 4.   Sedimentation profile of an extract from [³⁵S]sulfate-labeled cartilage on a 5–20% sucrose gradient under dissociative conditions. The bar graphs in the upper portion show the amount of binding of PGS by serum. The amount of binding of PGS in the cartilage extract before centrifugation is presented in the boxed-in bar graph. The open portion of the bar represents the percentages bound by immune rabbit serum; the cross-hatched portion, the percentage bound by normal rabbit serum; and the dot, the corrected specific binding. Sedimentation is from right to left [from Sparks et al, 1980].

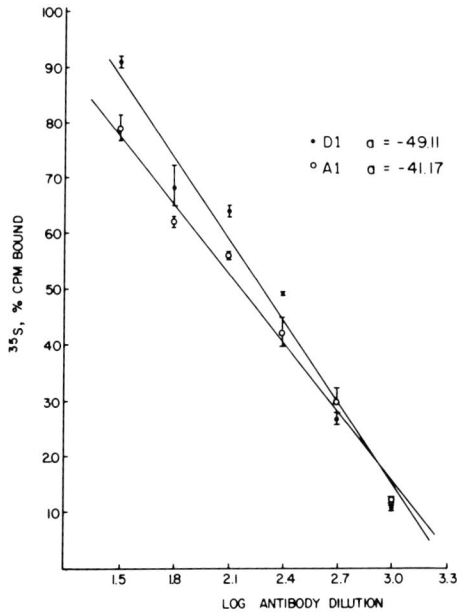

Fig. 5.   Regression curves of corrected specific binding of antigen plotted against the logarithm of antibody dilution. D1, bottom of a dissociation cesium chloride gradient, represents monomer. A1, bottom of an associative cesium chloride gradient, represents aggregates [from Sparks et al, 1980].

tibody cross-reacts with cartilage-specific PGS from the Japanese quail. The specificity of the antibody for cartilage PGS is indicated in Table I.

VI. CHONDROGENESIS IN MUTANTS

A. Nanomelia

1. Molecular analyses.   The first biochemical studies on the cartilage of nanomelic embryos was carried out by Mathews [1967]. The studies indicated that nanomelic cartilage contains only 10% of the ChS found in normal cartilage. The reduced quantities of ChS were shown to be a result of reduced synthesis and not of an increased degradation of that which is synthesized [Fraser and Goetinck, 1971]. When PGS is analyzed by sucrose density gradient centrifugation under conditions in which aggregates are dissociated, it is evident that the PGS-I seen in normal Meckel's and sternal cartilage extracts is absent or

TABLE I
Specificity of Binding of Immune Rabbit
Serum to Cartilage Proteoglycans

Antigens	% Bound by IRS corrected for NRS binding
1. [^{35}S]PGS	
a. Chick	
PGS (SC)	91.8
PGS (MC)	95.8
PGS (LC)	93.0
PGS (S)	8.2
PGS (LM)	2.6
b. Japanese quail	
PGS (SC)	96.0
2. [^{35}S]GAG	
a. Chick	
GAG (SC)	0.0
GAG (MC)	0.95
GAG (SF)	0.0
GAG (SF-xyl)	1.1

IRS, immune rabbit serum; NRS, normal
rabbit serum; SC, sternal cartilage; MC,
Meckel's cartilage; LC, limb cartilage; S, skin;
LM, limb mesenchyme; GAG,
glycosaminoglycan; SF, skin fibroblast; SF-
xyl, skin fibroblasts in presence of xylosides.

greatly reduced in the corresponding extracts of mutant organs (Fig. 6). The
different developmental histories of the cartilage precursor cells, therefore, have
no effect on the expression of the mutant gene in the differentiated tissues
[Palmoski and Goetinck, 1972; Pennypacker and Goetinck, 1976; McKeown
and Goetinck, 1979].

In order to find out which of the many steps in the biosynthesis of PGS might
be affected by the mutation, we took advantage of the observation that xylose
or a xyloside can stimulate ChS synthesis in the absence of core protein. This
approach allowed us to test indirectly the enzymes that synthesize ChS [Stearns
and Goetinck, 1979]. The results clearly indicate that nanomelic chondrocytes
have the potential to synthesize ChS, provided a suitable acceptor is present
(Fig. 7). Since these measurements do not take into account xylosyl transferase
activity, this enzyme was tested directly. The results of this analysis indicate
that this enzyme is also functional in the mutant (unpublished). All the results

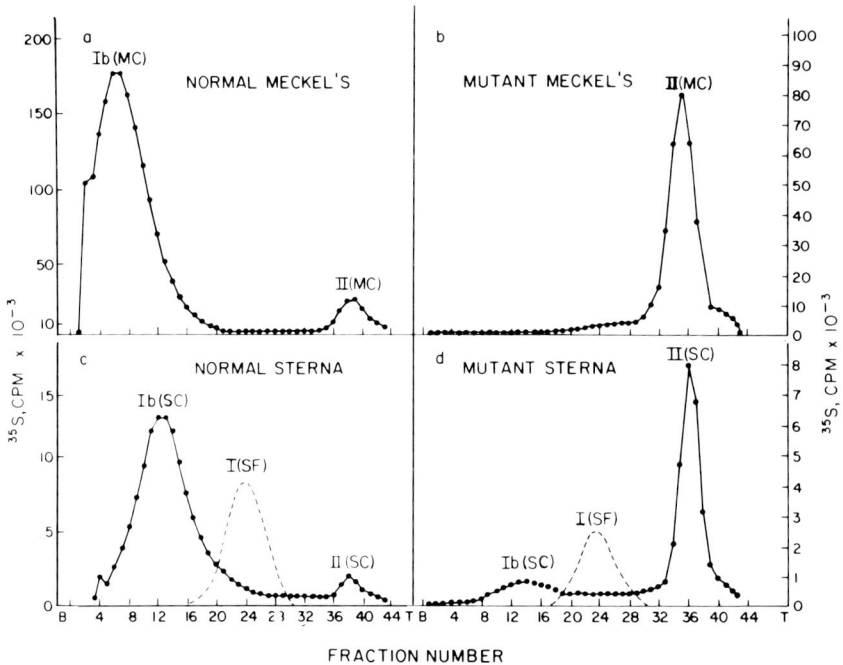

Fig. 6. Dissociative sucrose gradient profile of [^{35}S]sulfate-labeled PGS from extracts of the following tissues: a) normal Meckel's cartilage; b) nanomelic Meckel's cartilage; c) normal sternal cartilage; d) nanomelic sternal cartilage. The sedimentation of PGS-I of skin fibroblasts (SF) is indicated in c and d. Sedimentation is from right to left. [from McKeown and Goetinck, 1979].

with the nanomelic chondrocytes, therefore, indicate that the mutation affects the availability of PGS-I core protein.

Only cartilage is affected in the nanomelic mutant. The PGS synthesized by cultured skin fibroblasts of normal and nanomelic embryos is quantitatively and qualitatively indistinguishable [Goetinck and Royal, 1976; Goetinck and Pennypacker, 1977]. The major PGS of skin fibroblast [PGS(F)-I] has a lower molecular weight than the PGS-I of cartilage (Fig. 6). This information, as well as the fact that the nanomelic mutation affects the availability of the core proteins in cartilage and not in skin fibroblasts, suggests that the two PGSs (ie, skin fibroblast and cartilage) may differ in their core protein and therefore may represent different gene products.

Not only is the effect of the mutation restricted to cartilaginous tissues, but it only affects PGS of cartilage and not other cartilage-specific macromolecules,

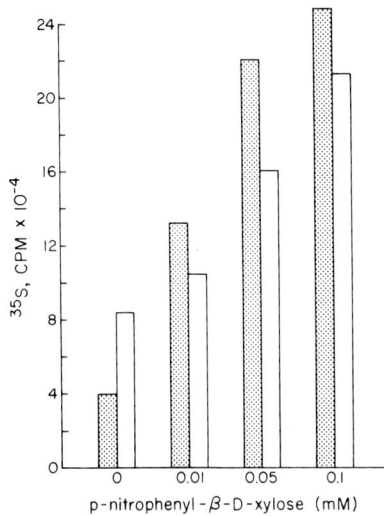

Fig. 7. Stimulation of [^{35}S]sulfate incorporation into chondroitin sulfate by the addition of p-nitrophenyl-β-D-xyloside to normal (white bar) and nanomelic (stippled bar) chondrocytes.

since the mutant cartilage contains normal levels of type II collagen [Pennypacker and Goetinck, 1976].

2. Cytological characterization.   Ultrastructural analyses of nanomelic cartilage reveal that the chondrocytes in the mutant are much more closely packed than the chondrocytes of normal cartilage (Fig. 8a, b). The cells of both phenotypes are indistinguishable from one another (Fig. 8c, d). However, an examination of the extracellular matrix reveals the absence of PGS granules in the mutant (Fig. 8e, f).

3. Immunochemical studies.   The antiserum generated against PGS was used to investigate the immunochemical properties of the greatly reduced quantities of PGS-I synthesized by nanomelic chondrocytes. This characterization was done by measuring the binding of PGS obtained by different analytical methods. The methods used were molecular sieve chromatography on controlled pore glass (CPG) [Lever and Goetinck, 1976] and sucrose density gradient centrifugation. When nanomelic and normal sternal extracts are chromatographed on CPG 1400, two peaks are obtained. In normals, most of the material chromatographs in the void volume [PGS(SC)-I], and very little is included [PGS(SC)-II] on the column. In contrast, only a small proportion of the mutant

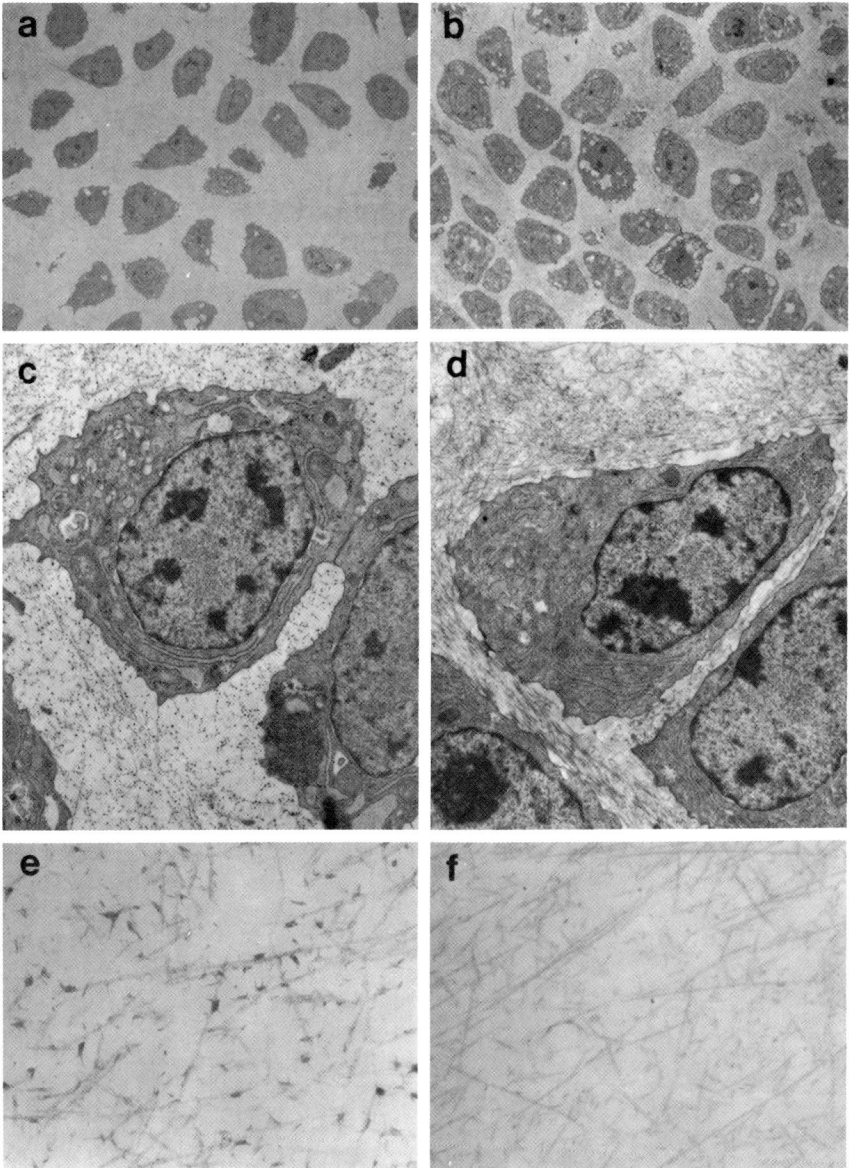

Fig. 8.   Ultrastructural features of normal (a, c, e) and nanomelic (b, d, f) sternal cartilage. The low-power electron micrographs reveal the higher density of the mutant (b) chondrocytes compared to normal (a). No differences are observed between normal (c) and mutant (d) chondrocytes, whereas the extracellular matrix of normal (e) cartilage possesses electron-dense matrix granules which are absent from the mutant (f) matrix.

extract chromatographs in the void volume (Fig. 9a). Antibody binding experiments on the CPG 1400 Vo fraction reveal that the serum binds more than 90% of the normal PGS-I but only 60% of the comparable mutant fraction (Fig. 9b) [McKeown et al, 1980].

In order to make a distinction between the monomeric and the aggregated forms of PGS(SC)-I, the CPG 1400 Vo material is chromotographed on CPG 2500, where the aggregate chromatographs in the void volume and the monomer is included. Most of the normal PGS(SC)-I is in the aggregated state, whereas only a small fraction of the mutant PGS(SC)-I is aggregated (Fig. 9c). Antibody binding measurements again indicate that only 60% of the mutant material can be bound, compared with more than 90% binding of the normal material (Fig. 9d) [McKeown et al, 1980].

When the CPG 1400 Vo material is examined on sucrose gradient under dissociative conditions, the sedimentation profiles seen in Figure 10a are ob-

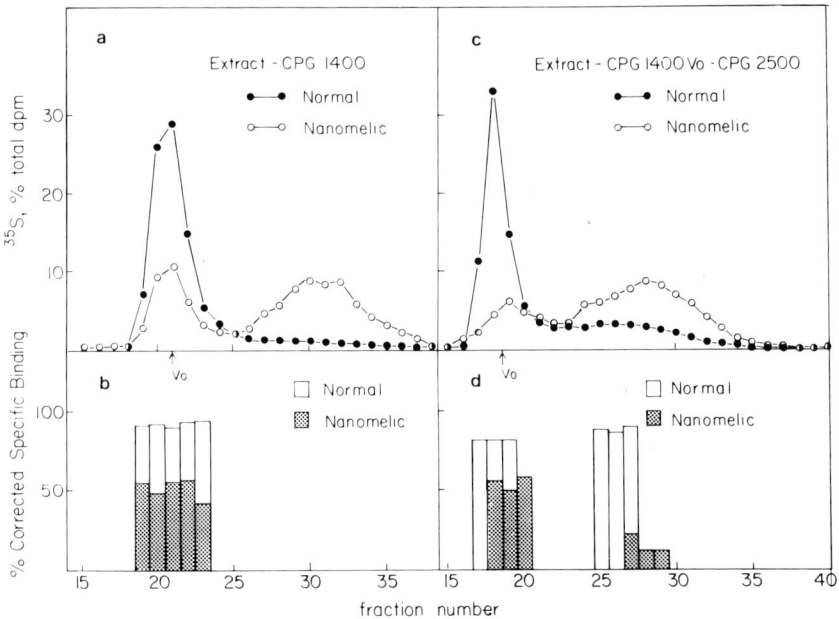

Fig. 9. Chromatographic and immunochemical analyses of PGS synthesized by normal and nanomelic sternal cartilage. a) Molecular sieve chromatography of sternal extracts on CPG 1400. PGS(SC)-I chromatographs in the void volume and PGS(SC)-II is included. b) Antibody binding of PGS(SC)-I. c) Molecular sieve chromatography of PGS(SC)-I on CPG 2500. PGS(SC)-I aggregates chromatograph in the void volume, and monomers are included on this column. d) Antibody binding of PGS(SC)-I, which chromatographs as aggregates and monomers.

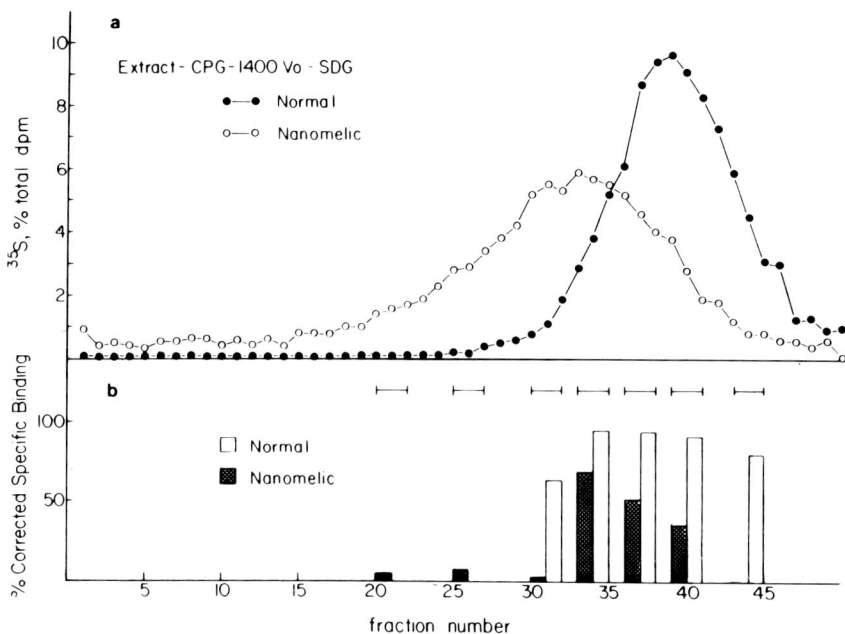

Fig. 10. Dissociative sucrose density gradient and immunochemical analysis of PGS(SC)-I of normal and nanomelic cartilage. a) Sucrose density gradient profiles. Sedimentation is from left to right. b) Antibody binding to designated fractions from the gradient.

tained. Nanomelic PGS(SC)-I has a slightly slower sedimentation rate than normal PGS(SC)-I, and the reduced binding with antibody is again evident (Fig. 10b) [McKeown et al, 1980].

Together the data on nanomelia indicate that the mutation results in greatly reduced quantities of core proteins of cartilage-specific PGS. The data demonstrate, furthermore, that the small quantity of PGS(SC)-I synthesized by nanomelic chondrocytes is not normal. This is clearly demonstrated by the reduction in antibody binding and the reduction in the amount of material capable of aggregating in the mutant extracts.

B. Micromelia-Abbott

1. Molecular analyses.   Accumulated PGS (measured as uronic acid per microgram of DNA) is lower in cartilaginous rudiments of the mm^A mutant than in normal comparable structures. PGS in the mutant cartilage is reduced to 30% in tibia and femora, to 40% in the sternum, and to 50% in Meckel's cartilage.

TABLE II
Incorporation of [^{35}S]Sulfate Into Sulfated
Proteoglycans by 13-Day Sterna

	N	Normal	mm^A
a) Control	3	624 ± 95[a]	188 ± 55[a,b]
b) + Puromycin	1	13	13
c) + Puromycin + Xyloside	5	405 ± 32	480 ± 88[b]

The means with similar superscripts are different from
each other at the following levels: [a]$P < 0.01$; [b]$P < 0.05$.

Synthesis of PGS (measured by the incorporation of radioactive precursors) is also reduced (Table II,a). The reduction in incorporation, however, is more severe in rudiments of a relatively early developmental stage than in the same rudiments at a later stage. Inhibition of protein synthesis by puromycin inhibits [^{35}S]sulfate incorporation (Table II,b). The addition of para-nitrophenyl-β-D-xyloside to the culture medium can stimulate the synthesis of sulfated glycos-aminoglycans by the mutant chondrocytes to normal levels (Table II,c). These results are interpreted to mean that the reduced levels of PGS may be the result of a reduced availability of xylosylated core protein for PGS. The PGS that is synthesized by the mutant chondrocytes is identical in sedimentation rate to the PGS synthesized by normal chondrocytes of the same rudiment [Quintner and Goetinck, 1980].

2. Cytological characterization. Ultrastructural analysis [Sawyer and Goetinck, 1980b] of mm^A cartilage reveals that there are more cells per unit area in the mutant than in the normal (Fig. 11a, b; Table III). Furthermore, the number of extracellular matrix granules of the mutant is greatly reduced compared to normal (Table III). The reduction in granule number is in close agreement with the reduction of PGS content. The mutant chondrocytes show less scalloping along their cell surface than do normal chondrocytes. The rough endoplasmic reticulum of the mutant resembles that found in normal chondrocytes, and the Golgi complex is well developed and consists of stacks of cisternae and small vesicles. However, the large Golgi vacuoles seen in normal chondrocytes are not a consistently observable feature in the mutant. The lack of large vacuoles may be related to the lack of extensive scalloping of the cell surface, since Seegmiller et al [1976] have suggested that scalloping may be a reflection of the exocytosis of vacuolar content.

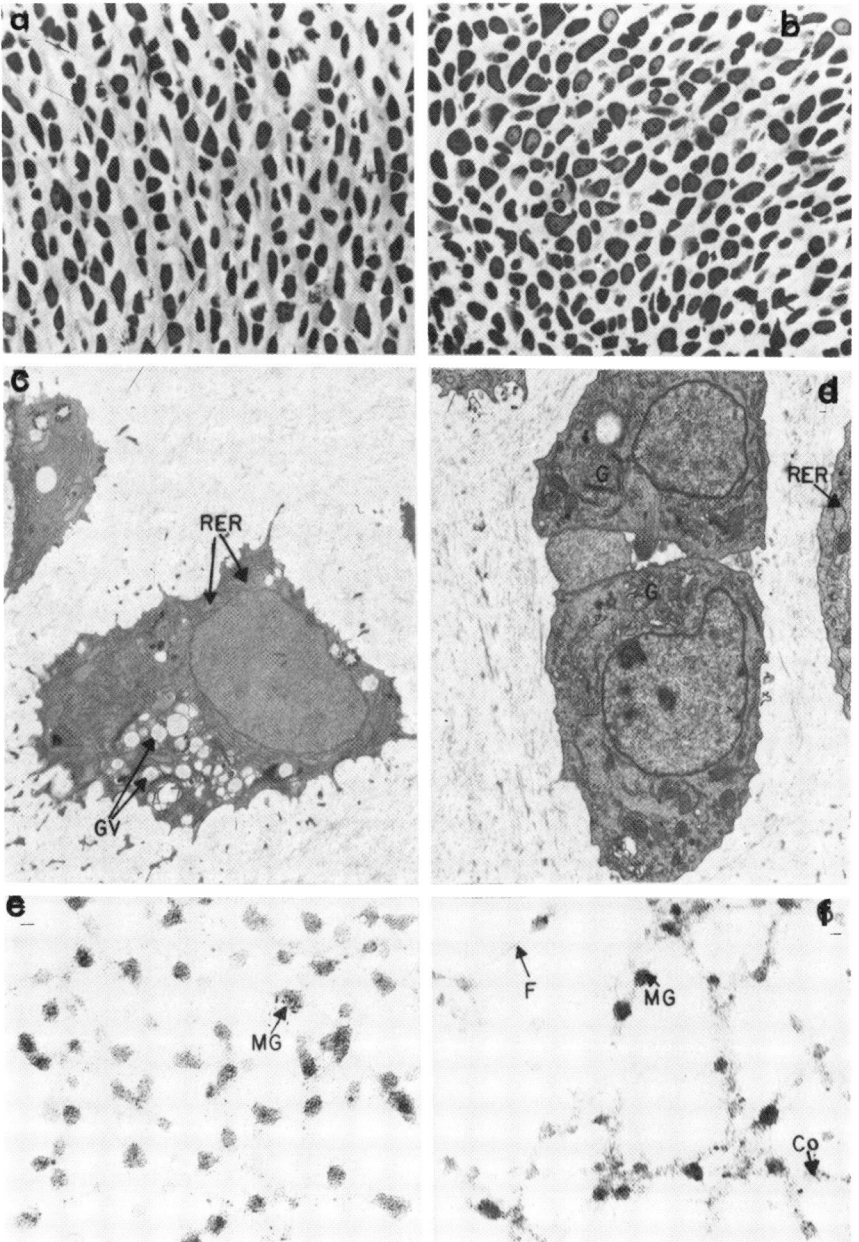

Fig. 11. Cytological analysis of normal (a, c, e) and micromelia-Abbott (b, d, f) cartilages. a) Low magnification picture, which reveals the higher cell density of the mutant (b) cartilage. The mutant chondrocytes (d) do not reveal an extension or scalloping of their cell surface as normal (c), and the mutant (f) has fewer matrix granules per unit area than the normal (e) [from Sawyer and Goetinck, 1980b].

TABLE III
Cell Density and Matrix Granule Content of
Normal and mm^A Cartilage

	Normal	mm^A
Ten-day tibia:		
Cells/unit area	5.3	12.3
Matrix granules/unit area	0.54	0.29
Ten-day Meckel's cartilage:		
Cells/unit area	5.2	7.3
Matrix granules/unit area	0.97	0.46

3. Immunochemical studies. Only preliminary immunochemical studies have been carried out with mm^A PGS-I. So far no obvious differences have been observed between mutant and normal.

VII. STUDIES ON PRECARTILAGINOUS TISSUES IN THE NANOMELIC MUTANT

The results on nanomelia described above indicate that PGS synthesis is affected in the mutant cartilage and not in noncartilaginous tissue. What is the situation in mutant precartilaginous tissues, and does the process of cartilage differentiation in the mutant differ from normal? Answering these questions presents special methodological problems. The mutant embryos are homozygous for a lethal gene and are, therefore, obtained from matings between heterozygous parents. Only 25% of the embryos resulting from such matings are expected to be mutant, and the identification of the embryos as mutant or normal can only be made with certainty at seven to eight days, when cartilage differentiation has already begun.

In order to study precartilaginous tissue in mutants, we have adapted the micromass culture method of Solursh et al [1978] to single embryos. In this culture system 2 × 10^5 limb bud cells are plated on a culture dish in 10 μl of culture medium. The cells are allowed to attach to the dish, and after two hours, the dish is flooded with culture medium. Under these conditions the limb bud cells undergo chondrogenesis. A total of seven to eight micromass cultures can be obtained from a single embryo. Two micromass cultures are established on one dish and cultured for six days for the phenotypic identification of the donor

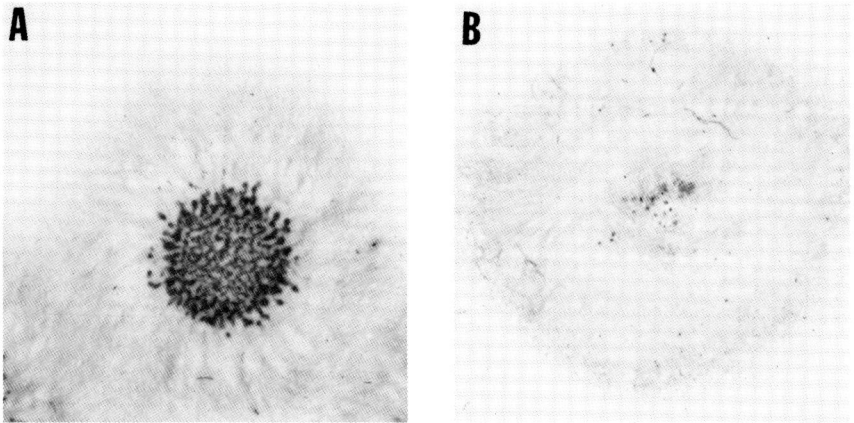

Fig. 12. Alcian green staining of micromass cultures of limb bud cells derived from single normal (A) and nanomelic (B) four-day-old embryos. The cells were maintained in culture for six days. Although the mutant cultures do not stain, they form cartilage nodules with an abnormal matrix [from Sawyer and Goetinck, 1980a].

embryo (Fig. 12). The remainder of the cultures are used for experimental analysis at any desired time point during the process of chondrogenesis.

Biochemical and morphological analyses have been done on normal and nanomelic micromass cultures after one, three, and six days of culture. This length of time covers the period of transition from undifferentiated mesenchyme to the formation of cartilaginous nodules. The results indicate that, at one day, there are no differences between normal and nanomelia in the type of PGS synthesized. At three days a difference is seen between the two genotypes when a more rapidly sedimenting peak appears in the PGS profile of normal but not of mutant. The difference is even more striking in six-day cultures (Fig. 13). Scanning electron microscopic analysis of the culture reveals no differences in the first and third days. At six days, however, a clear difference is evident in the ultrastructure of the extracellular matrix (Fig. 14). The mutant clearly lacks the matrix granules associated with the fibers [Sawyer and Goetinck, 1980a].

VIII. CONCLUDING REMARKS

This manuscript represents a progress report of work that was in fact begun by Dr. Walter Landauer. In our studies on chondrogenesis we have combined the methods of genetics, embryology, and biochemistry. This is the combination which Dr. Landauer predicted would lead ultimately to the understanding of

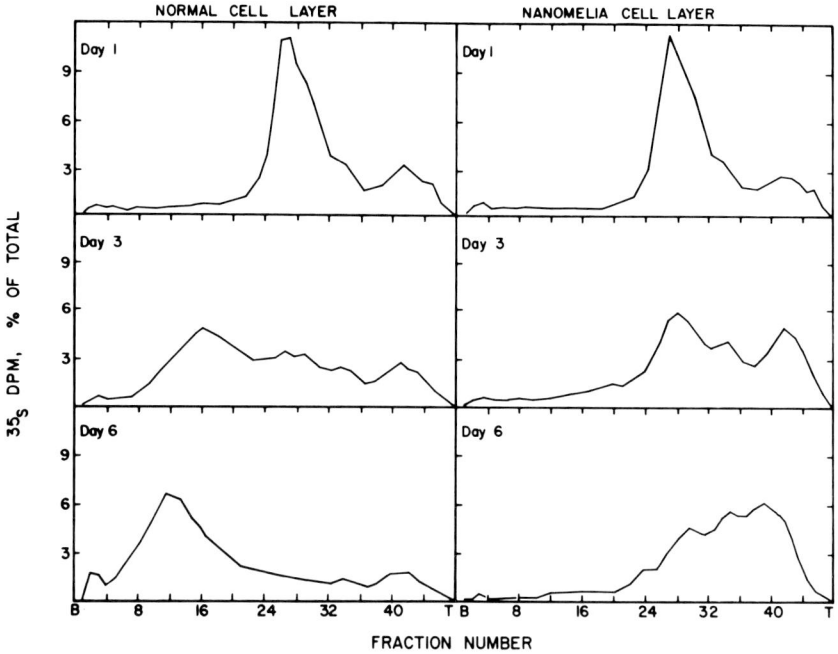

Fig. 13. Dissociative sucrose density gradient profiles of [³⁵S]sulfate-labeled PGS extracted from normal and nanomelic cell layers after one, three, and six days as micromass cultures. Direction of sedimentation is from right to left. Essentially the same profiles are obtained from the PGS in the medium [from Sawyer and Goetinck, 1980a].

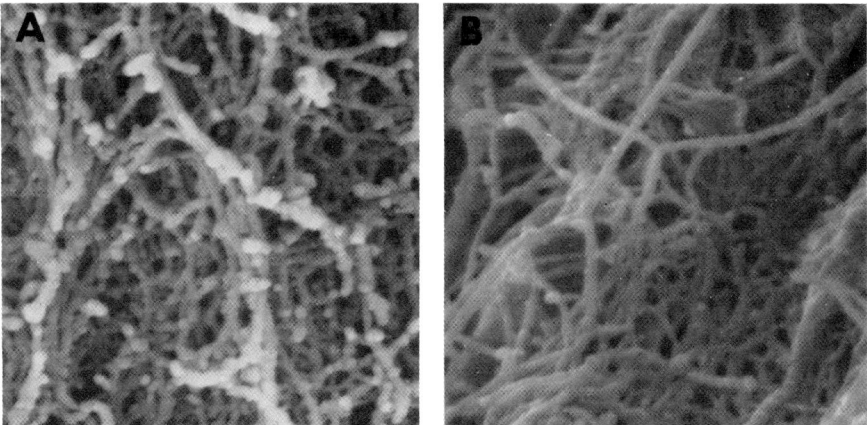

Fig. 14. Scanning electron micrograph of the extracellular matrix of normal (A) and nanomelic (B) six-day micromass culture [from Sawyer and Goetinck, 1980a].

gene action in development. Although some progress has been made toward these goals, we should remind ourselves of the opening statement of Dr. Landauer's address to the New York Academy of Sciences when he said: "It is important that we should not deceive ourselves. Our knowledge concerning the hereditary forces governing normal embryonic development, of the chick or any other vertebrate, is practically nil. This lack of information about the primary physiological function of genes in normal development should, of course, be a spur rather than a hindrance to future efforts" [Landauer, 1952]. For having enriched our scientific endeavors, all of us are greatly indebted to Dr. Walter Landauer. We dedicate this paper to his memory.

ACKNOWLEDGMENTS

The work reported here has been supported by grants GB 36790 from the National Science Foundation and HD 09174 from the National Institute of Child Health and Human Development. This paper is Scientific Contribution No. 859 of the Storrs Agricultural Experiment Station, The University of Connecticut.

REFERENCES

De Luca S, Lohmander LS, Nilsson B, Hascall VC, Caplan Al (1980). J Biol Chem 255:125–154.
Dorfman A, Ventrel BM, Schwartz NB (1980). Current Topics Dev Biol 14:169–198.
Farr RS (1958). J Infect Dis 103:239–262.
Fell HB (1939). Proc R Soc Lond B 229:407–463.
Fell HB, Canti RB (1934). Proc R Soc Lond B 116:316–349.
Fraser RA, Goetinck PF (1971). Biochem Biophys Res Comman 43:494–503.
Goetinck PF, Pennypacker JP (1977). In "Vertebrate Limb and Somite Morphogenesis." (DA Ede, JR Hinchliffe, and M Balls, eds.), pp 139–159. Cambridge University Press, London.
Goetinck PF, Royal PD (1976). J Gen Physiol 68:6a.
Hamburger V, Hamilton HL (1951). J. Morphol 28:49–92.
Hascall VC (1977). J Supramol Struct 7:101–120.
Ho PL, Levitt D, Dorfman A (1977). Dev Biol 55:233–243.
Johnston MC (1966). Anat Rec 156:130–143.
Kimata K, Okayama M, Oohira A, Suzuki S (1974). J Biol Chem 249:1646–1653.
Landauer W (1952). Ann NY Acad Sci 55:173–176.
Landauer W (1965a). J Hered 56:131–138.
Landauer W (1965b). J Exp Zool 160:345–354.
Le Lièvre C (1974). J Embryol Exp Morphol 31:453–477.
Le Lièvre C, Le Douarin NM (1975). J Embryol Exp Morphol 34:125–154.
Lever PL, Goetinck PF (1976). Anal Biochem 75:67–76.
Linsenmeyer TF (1974). Dev Biol 40:372–377.
Linsenmeyer TF, Smith GN Jr (1975). In "Extracelluler Matrix Influences on Gene Expression." (HC Slavkin and RC Gruelich, eds.), pp 303–310. Academic Press, New York.

Linsenmeyer TF, Toole BP, Trelstad RL (1973). Dev Biol 35:232–239.
Lohmander LS, De Luca S, Nilsson B, Hascall VC, Caputo CB, Kimura JK, Heinegård D (1980). J Biol Chem 255:6084–6081.
Mathews NB (1967). Nature 213:1255–1256.
McKeown PJ, Goetinck PF (1979). Dev Biol 71:203–215.
McKeown PJ, Sparks KJ, Goetinck PF (1980). Fed Proc 39:1638.
Miller EJ (1977). In "Cell and Tissue Interactions." (JW Lash and MM Burger, eds.), pp 71–86. Raven Press, New York.
Miller EJ, Matukas VJ (1969). Proc Natl Acad Sci USA 64:1264–1268.
Palmoski MJ, Goetinck PF (1972). Proc Natl Acad Sci USA 69:3385–3388.
Pennypacker JP, Goetinck PF (1976). Dev Biol 50:35–47.
Quintner MI, Goetinck PF (1980). Dev Genet (in press).
Sawyer LM, Goetinck PF (1980a). J Exp Zool (in press).
Sawyer LM, Goetinck PF (1980b). in preparation.
Seegmiller R, Ferguson CC, Sheldon H (1976). J Ultrastruct Res 38:288–301.
Solursh M, Ahrens PB, Reiter R (1978). In Vitro 14:51–61.
Sparks KJ, Lever PL, Goetinck PF (1980). Arch Biochem Biophys 199:577–590.
Stearns K, Goetinck PF (1979). J Cell Physiol 100:33–38.
Trelstad RL, Kang AH, Igarski S, Gross J (1970). Biochemistry 9:4993–4998.
von der Mark H, von der Mark K, Gay S (1976). Dev Biol 48:237–249.

The Molecular Control of Muscle and Cartilage Development

Arnold I. Caplan

Developmental Biology Center, Biology Department, Case Western Reserve
University, Cleveland, Ohio 44106

Levels of Genetic Control in Development, pages 37–68

I. INTRODUCTION

Cellular development proceeds via a progression of discrete decisional events as a multicellular organism develops. These decisional events eventually culminate in the commitment of individual cells or groups of cells into highly specialized, distinctive phenotypes with the subsequent phase of phenotypic expression in which unique macromolecules are assembled into the specialized arrays necessary for the functioning of that particular phenotype. The emphasis of this chapter is on these two distinct phases of cellular development: the commitment event and the expressional history of a committed cell.

It is important to recognize that each of these individual events does not occur in an instantaneous manner, but rather in a highly programed and time-dependent progression. The general program that governs these developmental transitions reflects the evolutionary and genetic history and bias of the organism. The individual cellular events not only reflect this evolutionary and genetic bias but also seem to be controlled by the complex, multivariable, and individual environments in which developing cells are situated. The purpose of this chapter is to attempt to describe at least one of the primary factors governing this highly specific commitment process, and also to describe the expressional history of a committed cell, of a chondrocyte. In both cases, it is clear that once cells progress through discrete stages of their developmental program (either commitment or expressional stages), then these cells seem to be incapable of back-tracking through a previous decisional or expressional state. The control of this forward directionality of development is obviously vested in the genetic program, of which we know preciously little.

II. THE DEVELOPING CHICK LIMB

For a variety of practical and conceptual reasons, the developing chick limb provides a superb vehicle for experimentation into aspects of phenotypic commitment and expression. Primarily, the limb is accessible as an external appendage to experimental manipulation, and its development can be timed with a high degree of accuracy [Hamburger and Hamilton, 1951; Saunders, 1948].

The origin of limb cells is graphically depicted in Figure 1, where the cross-hatched areas indicate the mesenchymal location of cells in the *limb field*. By day 1 of chick embryo development, the single group of fated limb cells bifurcates into fore- and hind limbfields. At this point, a dramatic and powerful commitment takes place, in which these individual groups of cells become committed to a specific form; ie, cells are committed to either fore- or hindlimb morphology. Experimentation indicates that these fore- and hindlimb groupings of cells may not be interchanged for one another although the eventual phenotypes, muscle,

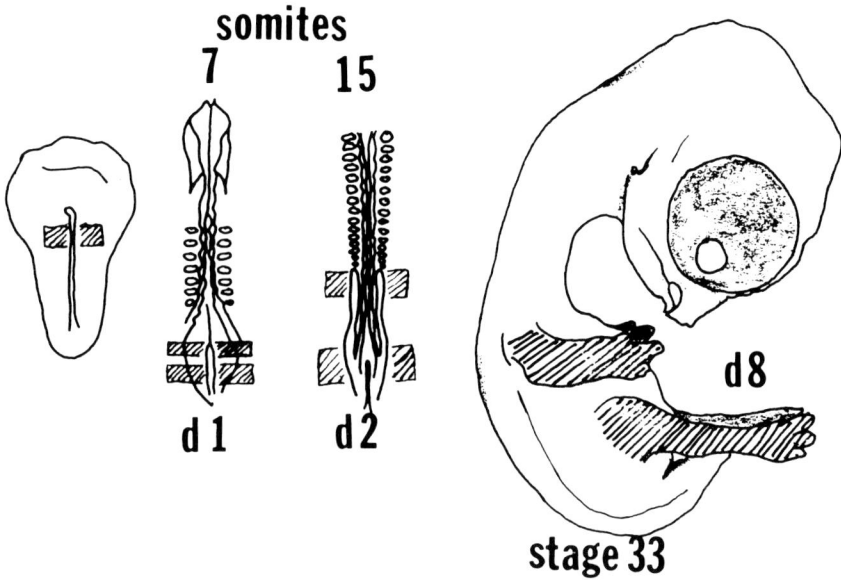

Fig. 1. Chick embryo limb field: Primitive streak, one day (d1), two day (d2), and eight day (d8) embryos are pictured, with the shaded area representing wing and leg limb fields and their eventual maturation into limbs.

cartilage, and bone, are not yet determined [Rudnick, 1945; Wolff, 1936]. The cells of the flank mesenchyme originate from the mesenchymal mass, which eventually gives rise to somite, lateral plate, and splanchnic mesenchyme. This mass of mesenchyme eventually sends out two sheets: one that will become lateral plate mesenchyme and one that will populate the internal membranes (splanchnic mesenchymal areas) of the embryo; this original mass will also eventually segmentalize into individual somites. Recent experimentation also indicates that some cells from the somite actually migrate into the limb field in the flank mesenchyme and may exclusively give rise to muscle [Christ et al, 1977; Kieny and Chevallier, 1979]. Thus, before the development of an individual limb bud, the cells of the limb field have become committed to a particular form, although by a variety of experimental tests they seem to be pluripotent for various phenotypes [see Zwilling, 1968].

The series of events involved with commitment of such mesenchymal cells into these particular limb phenotypes and their subsequent expressional history are of experimental interest in this chapter. The work described in the pages that follow is predicated on the assumption that the cells of the limb mesenchyme are uncommitted for phenotype and become committed at a discrete time window

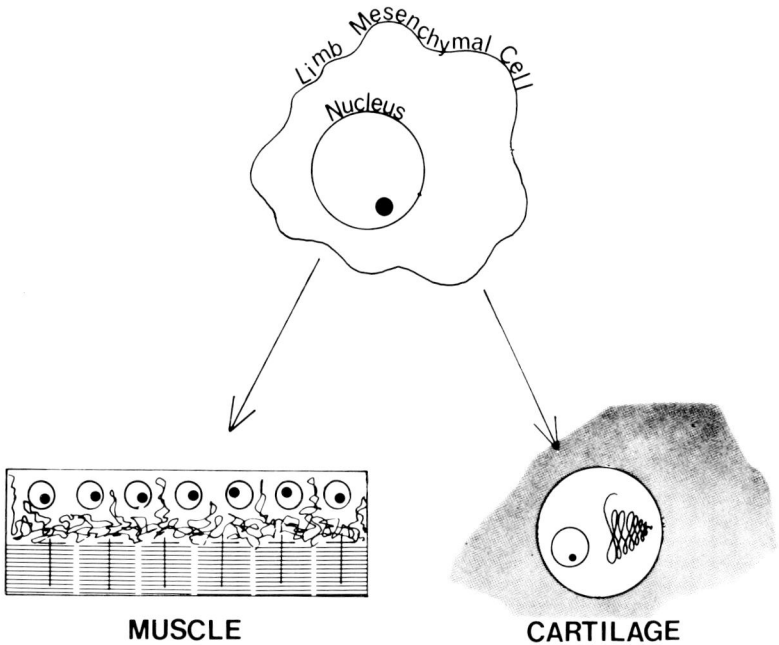

Fig. 2. Limb mesenchymal cell can develop into muscle or cartilage phenotypes.

of development. Experimentation that is consistent with the pluripotentiality of such limb cells has been reported in detail elsewhere [Zwilling, 1968; Finch and Zwilling, 1971; Searls and Janners, 1969] and will not be reviewed here. It is sufficient to say that such experimentation is not unequivocal and that sufficient controversy now exists with regard to the commitment and development of the muscle phenotype. Thus, this is still a working assumption and, as will be discussed later, direct experimental proof of such assumptions is required.

With this proviso, Figure 2 pictorially emphasizes the primary focus of the discussion. What controls the choice of individual mesenchymal cells into particular phenotypes and governs their subsequent expressional progression? It should be clear from the outset that, although muscle and cartilage cells have a number of cellular characteristics in common (they both have cell membranes, mitochondria, nuclear membranes, histones, glycolytic enzymes, etc), these phenotypes are as distinctive and different as any two phenotypes found in the body. In one case, muscle cells assemble a complicated intracellular macromolecular array involving the contractle apparatus; cells are multinucleated; and their early energy metabolism is predominantly aerobic. In the other case, chondrocytes primarily secrete large amounts of proteins and polysaccharides, which

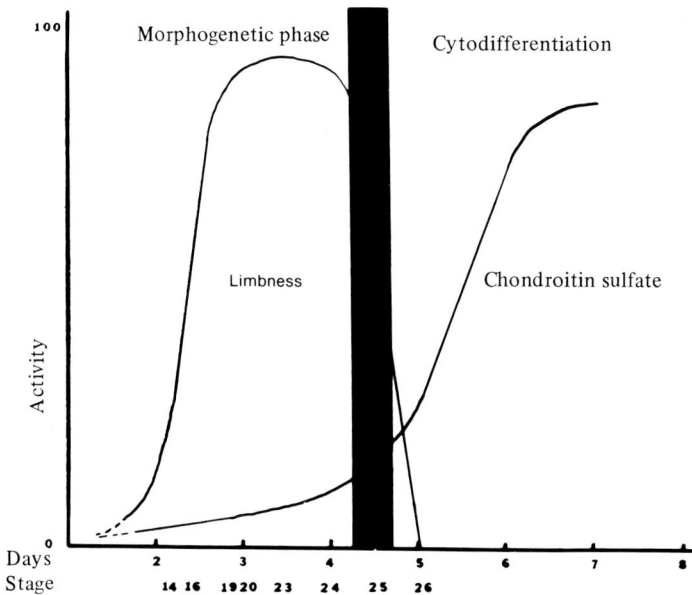

Fig. 3. Graphic summary of various observations in which Zwilling [1968] plotted the idealized transition between the pluripotential cell state capable of responding to apical ectodermal ridge and the loss of this capacity as the mesenchymal cells of the limb become cytodifferentiated and exhibit morphological and biochemical characteristics of committed and expressive cells. The black vertical bar has been added by this author to indicate the time window when limb mesenchymal cells become committed to specific phenotypes.

organize around the cell to form a unique extracellular matrix; these cells are mononucleated and are highly glycolytic (anaerobic) in their energy metabolism.

Summarizing early phenomenological information with regard to the timing of decisional events involved in limb development, Zwilling [1968] drew the figure represented in Figure 3. He concluded that limb mesenchymal cell development could be experimentally characterized by two discrete phases: a morphogenetic phase and a cytodifferentiation phase. In the morphogenetic phase, the cells were experimentally interchangeable with regard to their position within the limb and were capable of responding to the inductive interactions with the apical ectodermal ridge of the ectodermal jacket to provide a growing appendage. As these cells lost their ability to respond to the inductive involvement of the apical ectodermal ridge, they seemed to augment or emphasize highly specialized functions summarized by the term "cytodifferentiation"; in the case of the developing chondrocyte, the cells demonstrated the increased ability to synthesize specific chondroitin sulfate-rich molecules. The discrete time window (between stage 24 and 25) when cells seem to lose their pluripotentiality and initiate

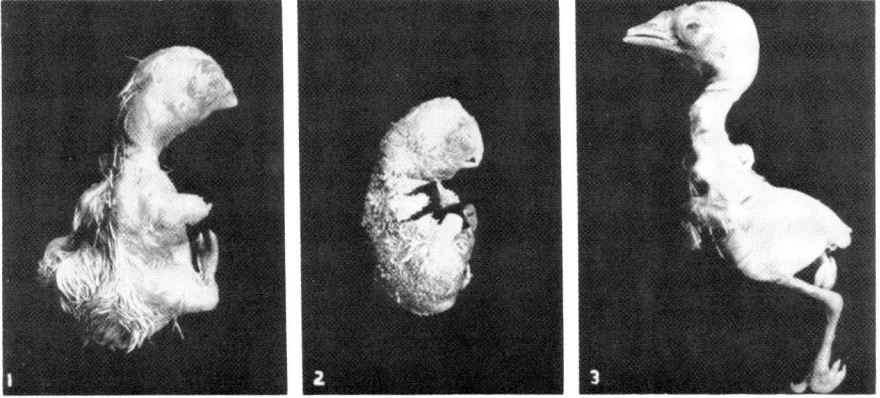

Fig. 4. Reproduction of Figures 1, 2, and 3 from Landauer [1957]. Figs. 1 and 2. Full-term embryos, treated at 96 hours with 10 μg 6-aminonicotinamide. Micromelia and parrot beak. Down partly removed from embryo shown in Fig. 1; down abnormal in embryo shown in Fig. 2. Fig. 3 Full-term embryo, treated at 96 hours with 0.75 mg 3-acetylpyridine. Note almost complete lack of musculature along tibiatarae. Down removed.

cytodifferentiation is represented by the black vertical bar in Figure 3. To restate the question addressed here: What influences a mesenchymal cell during this discrete time window to decide to become cartilage or muscle phenotypes? The body of information provided here experimentally addresses this question.

III. HISTORICAL PERSPECTIVE: WORK OF WALTER LANDAUER

In the 1950s Walter Landauer and his collaborators reported a variety of experiments in which particular chemicals were introduced into developing chick eggs at 96 hours of development, with the subsequent characterization of their individual teratogenic effects [Landauer, 1948, 1954, 1957; Landauer and Clark, 1962; Landauer and Wakasugi, 1967; Landauer and Sopher, 1970]. The theme of Landauer's experimentation was to not only define which compounds caused what family of teratogenic effects but also to identify those naturally occurring antagonists of such teratology. Of particular interest were his detailed studies of two groups of such teratogenic agents: one group caused muscle malformations, while the other group caused cartilage malformations. Figure 4 is a reproduction from Landauer's 1957 report describing two such teratogens, 3-acetylpyridine, which causes muscle malformations, and 6-aminonicotinamide, which causes cartilage malformations. Of particular interest in this report and others by Landauer and his collaborators [Landauer, 1948, 1954, 1957; Landauer

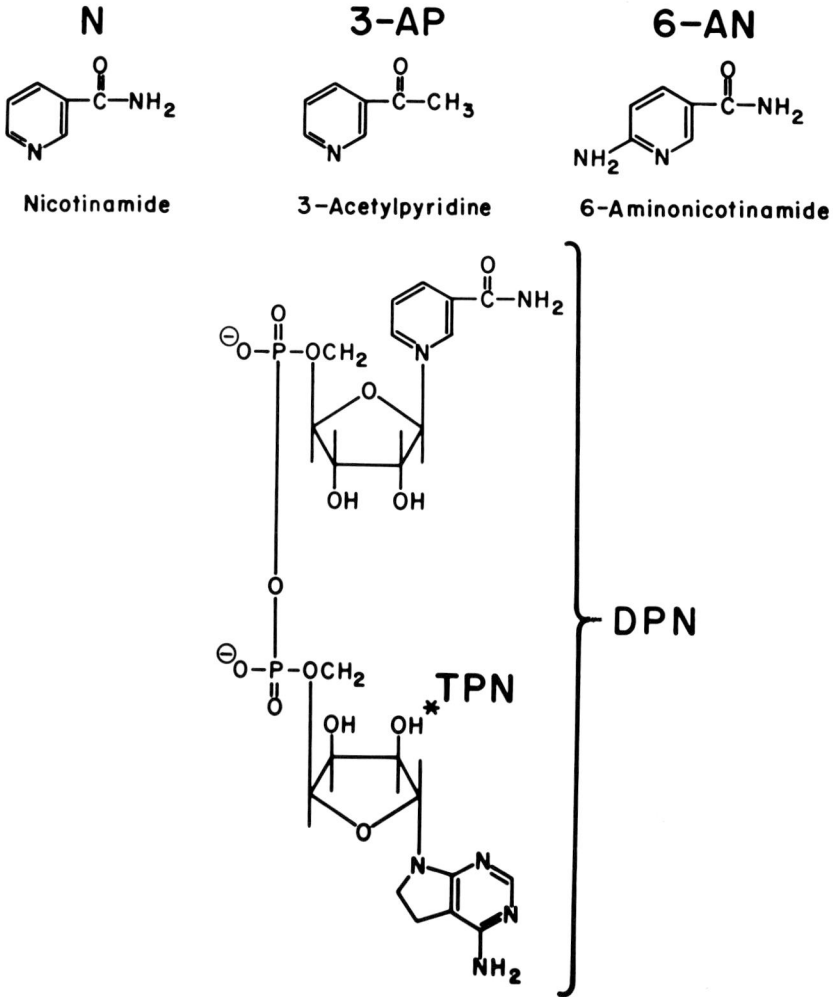

Fig. 5. Chemical formulas for nicotinamide (N), 3-acetylpyridine (3-AP), 6-aminonicotinamide (6-AN), nicotinamide adenine dinucleotide (DPN or NAD), and nicotinamide adenine dinucleotide phosphate (TPN or NADP).

and Clark, 1962; Landauer and Wakasugi, 1967; Landauer and Sopher, 1970] is the observation that a single compound, nicotinamide, antagonizes these two groups of malformations; nicotinamide in sufficient quantities will completely protect against *both* types of teratogenic effects.

Figure 5 shows the molecular structures of nicotinamide, 3-acetylpyridine, and 6-aminonicotinamide. It is clear that the chemistry of these teratogenic effects is related to the chemistry of NAD, since nicotinamide is either directly incorporated into NAD or is excreted from cells and organisms as the N-methyl nicotinamide. Additionally, at this time other workers had clearly established that the extracellular levels of nicotinamide directly control the internal pool sizes of NAD [Handler and Klein, 1942a, b; Kaplan et al, 1954; Oide, 1958]. It was also known that NAD was unable to move across cellular or mitochondrial membranes, although nicotinamide easily crosses membranes thus providing a mechanism by which the extracellular concentrations of nicotinamide directly control the internal concentration of NAD. Landauer's elegant analysis of the teratogenic effects of 3-acetylpyridine and 6-aminonicotinamide indicated to him that these teratogens were specifically affecting the cell's energy metabolism pathways and therefore were affecting the mesenchymal cell's developmental pathways. Such stimulating hypothetical considerations motivated me to attempt to establish the precise molecular basis for the action of these two families of teratogens and led me to suspect that nicotinamide itself may play some central role in the decision-making process by which limb mesenchymal cells commit into muscle and cartilage phenotypes.

It was through a fortunate series of circumstances as a postdoctoral fellow in the laboratory of Nathan O. Kaplan that I attempted to embark upon such experimental analysis. The first experiment I attempted involved setting up myoblast cultures as had been reported by Hauschka and Konigsberg [Konigsberg, 1960; Hauschka and Konigsberg, 1966] and to expose such cultures to the muscle teratogen, 3-acetylpyridine, and nicotinamide in an attempt to delineate a molecular mechanism of action of these molecules. With this in mind, Nathan Kaplan introduced me to Edgar Zwilling and arranged for me to do the tissue culture experimentation in Zwilling's quarters at Brandeis. Although Landauer and Zwilling had been close colleagues at University of Connecticut [Zwilling and DeBell, 1950; Zwilling, 1948, 1949; Landauer, 1948] and Zwilling and Kaplan had been close colleagues at Brandeis, it was my interest in the molecular details of Landauer's earlier teratological studies which brought the work and interests of these three experimenters together. It wasn't long after coming in contact with Edgar Zwilling that I was able to provide the experimental bridge between the earlier work and hypotheses put forward by Zwilling with those of Walter Landauer.* Since Nathan Kaplan had always been interested in the role of nicotinamide and NAD in cellular events, he was highly supportive of this avenue of experimentation.

---

*Landauer himself provided the intellectual bridge, and this was long recognized by both Zwilling and Kaplan, but no one worked on these problems until I was moved by Landauer's comment: "These observations indicated clearly that nicotinamide, presumably as part of phosphopyridine nucleotides, plays an important role in various steps of morphogenesis of the embryo. Hence, it seemed of interest to inquire into the effects on chick development of compounds which, from independent evidence, are known to be nicotinamide competitors" [Landauer, 1957].

IV. TISSUE CULTURE OF UNCOMMITTED LIMB
MESENCHYMAL CELLS

Using 3-acetylpyridine and nicotinamide, I was able to demonstrate in Konigsberg-Hauschka developing myoblast cultures that 3-acetylpyridine specifically affected myoblasts and had little or no affect on fusion and myotube maturation [Caplan et al, 1968]. As a control experiment, Zwilling and I decided to put stage 24 chick limb mesenchymal cells into culture and to ascertain the affects of 3-acetylpyridine on these cells directly. We chose stage 24 limb buds as our starting material since that represented the stage of development just prior to commitment of these mesenchymal cells into muscle or cartilage phenotypes. We chose a plating density of cells for our initial experiment and a level of 3-acetylpyridine that was based on a long series of complicated and somewhat reasonable but, retrospectively, very "lucky" assumptions. What resulted was a culture of limb mesenchymal cells that had 20–30% of the cells in the untreated controls exhibiting properties associated with differentiated chondrocytes, 10–20% with differentiated myogenic elements, and the remainder as unidentifiable fibrogenic or connective tissue phenotypic elements. When such cultures were exposed to high levels of 3-acetylpyridine during the first days of culture, all of the myogenic elements that should have developed did not do so with the startling and rather spectacular potentiation of chondrogenic development. Later analysis showed that under these conditions greater than 90% of the cells in the culture exhibited a chondrogenic phenotype [Caplan, 1970, 1972; Caplan and Stoolmiller, 1973]. With this observation, the emphasis of the experimentation shifted from studies designed to understand the mode of action of a muscle teratogen to an attempt to understand the potentiating effects of this in vivo muscle teratogen on the commitment and expression of chondrocytes in vitro.

Published data from my laboratory indicate that 3-acetylpyridine functions to potentiate chondrogenic expression by specifically inhibiting the entrance of nicotinamide into cells with the subsequent and dramatic lowering of the NAD pool size by one or two orders of magnitude [Caplan, 1970, 1972; Rosenberg and Caplan, 1974, 1975; Caplan and Rosenberg, 1975a, b]. This lowering occurs during the period when cells are committing to particular phenotypes such that the internal pool sizes of NAD are low, a condition that seems to be stimulatory for chondrogenic expression. Subsequent experimentation indicated that when the external levels of nicotinamide were raised, causing increased levels of NAD within the cells, then chondrogenic expression was specifically inhibited while myogenic expression was maintained and/or enhanced. This correlation of high NAD levels with muscle development and of low NAD levels with cartilage development was also obtained when measurements were done on in vivo material in the whole limb bud [Caplan, 1970, 1972; Rosenberg and Caplan, 1974, 1975; Caplan and Rosenberg, 1975a, b]. Thus, as re-represented in Figure 6,

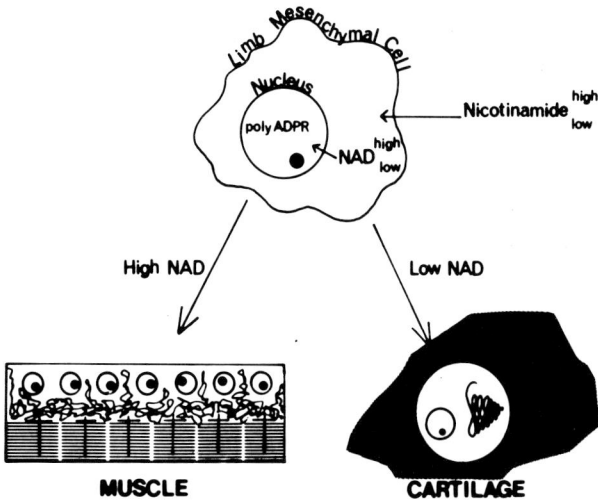

Fig. 6. Development of limb mesenchymal cell into muscle and cartilage as influenced by external nicotinamide levels, cellular NAD levels and the nuclear synthesis of poly (ADPribose).

extracellular concentrations of nicotinamide seem to control internal pool sizes of NAD resulting in preferential commitment and expression of either muscle or cartilage in conditions of high NAD versus low NAD, respectively.

A. Density-Dependent Commitment and Differentiation of Limb Mesenchymal Cells

Since nicotinamide levels are determined by extracellular conditions, we have systematically investigated the effect of variation in original seeding density on phenotypic expression as a vehicle for varying nutrient accessibility. Cells that are plated at low densities tend to spread out and have a high surface-to-volume ratio; in this case the accessibility to nutrients is maximized and the competition for diffusing nutrients above individual cells is minimized. At the other extreme, under very high initial seeding densities, cells either are in a multilayer or are forced into a highly cuboidal morphology at the initiation of cultures. These cell-packing conditions, thus, provide a low surface area-to-volume ratio, and because there are so many more cells per unit surface area, an intense competition is established for nutrients in the tissue culture medium. As represented in Figure 7, cultures are established from the same isolate of single cells from stage 24 limb bud; such cultures are seeded at low (1×10^6 cells/35 mm dish), medium (2×10^6 cells/35 mm dish), or high density (5×10^6 cells/35 mm dish). The observed phenotypic expression varies with the initial seeding density. Analytical

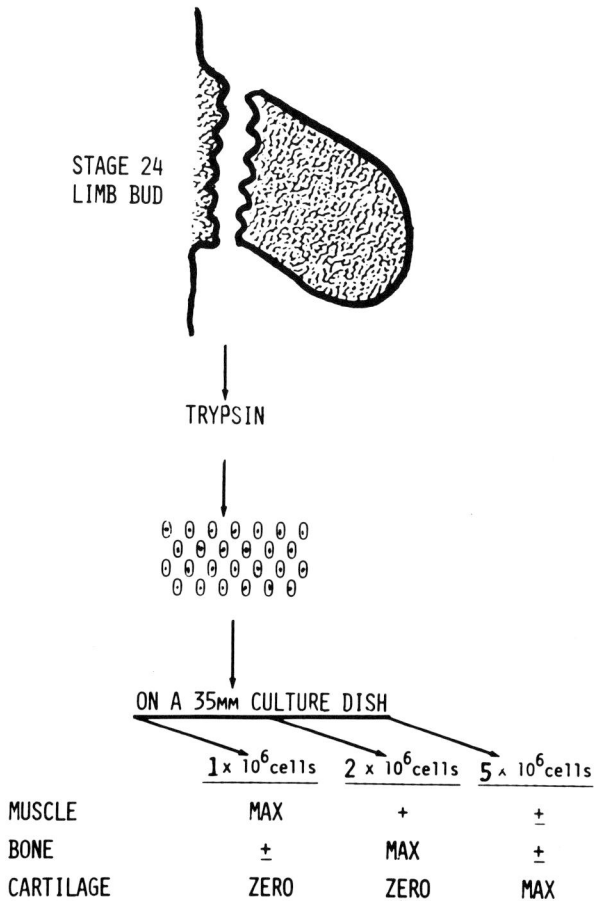

Fig. 7. Schematic representation of the preparation of stage 24 limb mesenchymal cells for culture and the differential phenotypic expression as a function of initial seeding density.

morphology reported elsewhere [Osdoby and Caplan, 1979, 1980] indicates that muscle development is maximal at low density, with absolutely zero chondrogenesis being observed until cells are plated above two million cells per 35 mm dish. Bone differentiation, seen as osteoblast and osteocyte morphology, is maximal in medium-density cultures while at high-density conditions, chondrogenesis dominates, with 50–85% of the cells identified as chondrocytes. Although we have argued that this density dependence reflects nutrient accessibility, it is probable that there are other important parameters involved in accentuating these different phenotypic properties.

B. An Assay for In Vitro Commitment

The question arises of exactly *when* following introduction of cells in the cultures do cells commit to a particular phenotypic pathway. Recent attempts to answer this question involve use of the above density-dependent differentiation pattern. Briefly, we have asked, "When do cells commit to a *non-chondrogenic* phenotype?" We have attempted to answer this question by initially plating cells at low density where they will never express a chondrogenic phentoype. Subsequently, at various times after plating of these low-density cultures, the cells are removed from the plate with trypsin, counted, and replated at high density under conditions we know are optimal for chondrogenic development. Our thought is that those cells that have not become committed to a non-chondrogenic pathway will differentiate into chondroytes under conditions of high density that are optimal for chondrogenic expression. Preliminary results [Yander, Cockrell and Caplan, unpublished] indicate that at six hours and 12 hours after cells are plated at low density, replating them at high density encourages the majority of cells to express a chondrogenic phenotype. Between 18 and 36 hours an ever-decreasing proportion of cells are able to express a chondrogenic phenotype when placed from low-density to high-density conditions. After 36 hours, such replated low-density cells seem to be incapable of committing or expression of chondrogenic phenotype. Thus, the commitment into particular phenotypic pathways seems to occur at sometime between 18 and 36 hours after cells are seeded into the culture.

V. THE RELATIONSHIP BETWEEN CYCLIC AMP AND NAD

Since we have established that the extracellular environment seems to be an important parameter in the commitment and expression of chick limb mesenchymal cells, it is natural to attempt to investigate other systems that are capable of responding to changes in extracellular environment which then subsequently affect the differentiation and development of these embryonic cells. One system that immediately comes to mind is the cyclic AMP (cAMP) system, which is well known to translate extracellular events into modifications of cell activities and intracellular pool sizes of cAMP. In addition, speculation has been put forth by McMahon [1974] arguing that cAMP levels are related to NAD and also are related to events associated with control aspects of limb development.

As pictured in Figure 8, the cellular cAMP levels seem to vary with the initial plating density of stage 24 chick limb mesenchymal cells [Youngman and Caplan, unpublished]. By day 1 in culture, the cells plated at high density (which will differentiate into chondrocytes) have very high internal levels of cAMP, whereas cells plated at low density have low levels of cAMP. These differences

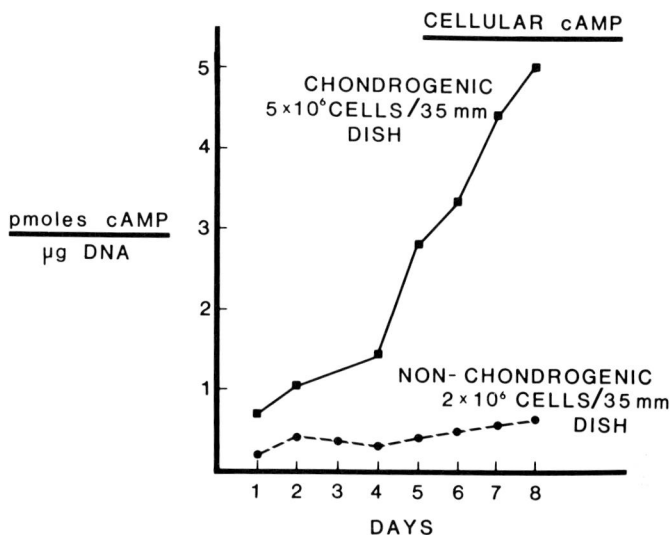

Fig. 8. Basal levels of cyclicAMP (cAMP) per unit cellular DNA when cells are initially seeded at "intermediate" density (2×10^6 cells/35 mm dish) or "high density" (5×10^6 cells/35 mm dish). The predominant phenotype as described in Figure 7 is osteoblast for intermediate density with no cartilage present; high-density cultures are predominantly (50–85%) chondrogenic, with very little bone or muscle cells.

are accentuated as the cells progress through their expressional histories, as seen in Figure 8. Thus, under conditions of high internal cAMP levels, chondrogenesis is enhanced, while low cellular cAMP levels seem to favor nonchondrogenic expression. This relationship is exactly the reciprocal of that seen for NAD, and it may be that cAMP and NAD are involved in a functionally reciprocal relationship. Since cAMP is referred to as a "second messenger" and since cAMP and NAD seem to be related to the differentiation pattern seen under these defined in vitro conditions, it may be reasonable to think of NAD as an informational molecule as well as an enzymatic cofactor.

VI. THE CONTROL OF NUTRIENT ACCESSIBILITY IN THE LIMB: PHENOTYPIC POSITIONAL INFORMATION

Observations on the cultured stage 24 limb mesenchymal cells indicated that exposure to different extracellular environments influenced the phenotypic commitment of these cells. These observations included variations in internal NAD levels as controlled by the extracellular concentrations of nicotinamide, the presence of modulators that affect cAMP levels, and the physical constraints involved

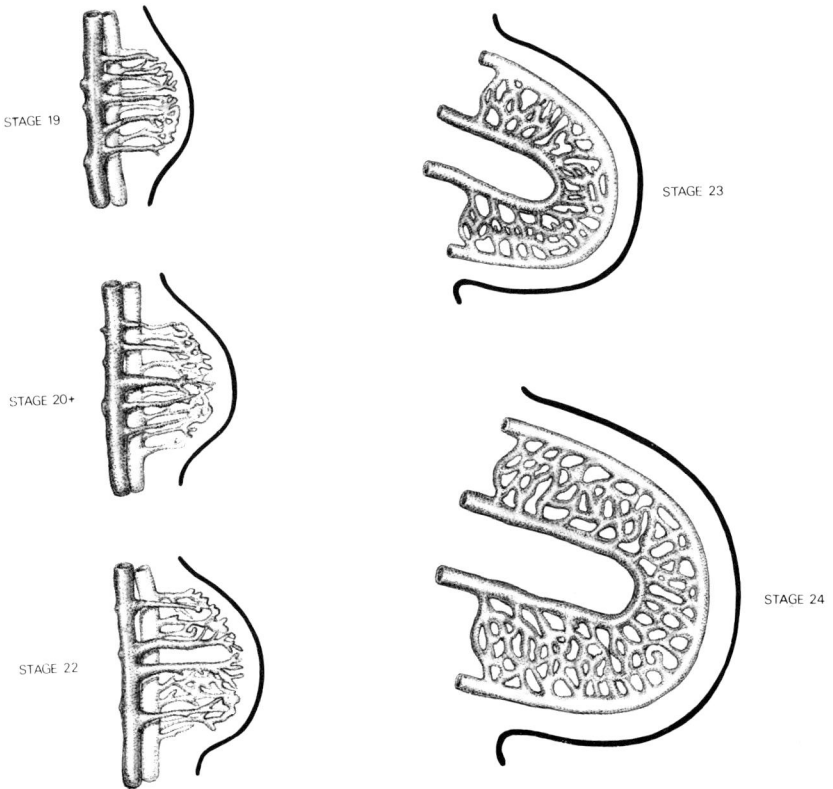

Fig. 9. Artist's schematic of the vascular pattern in the embryonic chick limb at various stages of development; drawings are made from limbs visualized from India ink injections published previously [Caplan and Koutroupas, 1973].

in packing density variations. These observations can be used to support the general hypothesis that there must be a physiological basis for differential extracellular environments in the developing limb buds. With this hypothesis in mind, Steve Koutroupas and I [1973] attempted to delineate the developmental progression of the embryonic chick vascular system. As pictured in Figure 9, the results of this investigation indicated that by stage 22 or 23, the central core area of the limb was totally avascular, while the major capillary sprouts were located in the peripheral, premyogenic areas. Thus, the development of the vascular system provides differential nutrient environments (core versus periphery) in the developing limb bud. As schematized in Figure 9, cells in the premyogenic area would be exposed to higher rates of nutrient flow than cells at the prechondrogenic core. It should be remembered that this compartmentalization of nutrient flow is established prior to the actual phenotypic commitment of these limb mesenchymal cells, which occurs after stage 24. Thus, one might tentatively conclude that the limb mesenchymal cells must be exposed to differential nutrient environment for a specific time interval prior to their eventual commitment into a particular phenotypic pathway.

Retrospectively, I believe this makes sense from a teratological viewpoint. If the embryo is exposed to a high and instantaneous dose of a teratogen, it must recover or perish. It is clear that the cells are not susceptible to single intense pulses of reagents, since their developmental program is geared for environmental cues, which must be relatively constant over large blocks of time (probably hours for chick embryos). Thus, molecules or reagents that are influential or that might serve to control developmental pathways must be exposed to cells over long periods; likewise, molecules that are detrimental to cells (teratological agents) must be exposed to cells over a relatively long period.

From the above, I would argue that the vascular systems set up two environmental compartments in which uniform cells of uniform history will be obliged to develop along different developmental pathways. The experiments of Searls and Janners [1969], Zwilling, and others [see Zwilling, 1968 for review] in which blocks of tissues were placed in reverse orientation or tissue was randomized must therefore be compatible with the reestablishment of this nutrient compartmentalization. In a speculative sense, the morphogenetic determinant for limb may indeed be those cellular components that structure vascular invasion and subsequent nutrient flow.

VII. CHROMOSOMAL PROTEIN MODIFICATION RELATED TO NAD AND COMMITMENT EVENTS

The question arises as to whether NAD pool sizes have their major effects on developmental pathways by controlling energy metabolism or by other means. From a variety of tests conducted in my laboratory, it would appear that shifts in NAD levels do not have causal effects on development because of their involvement in particular energy pathways. Recently, it has come to our attention that a unique nucleic acid polymer called poly(ADPribose) forms from the polymerization of NAD, with simultaneous excision of nicotinamide [Chambon et al, 1963; Harris, 1973; Okayama et al, 1976]. These reactions are schematized in Figure 10: a specific enzyme referred to as a poly(ADPribose) polymerase is responsible for the synthesis of this polymer and a specific enzyme referred to as the poly(ADPribose) glycohydrolase is responsible for cleaving the unique ribose–ribose linkage to degrade this polymer. The polymer itself is covalently bound to chromosomal proteins (both histone and non-histone acceptor proteins have been identified). Thus, the two specific enzymes for the synthesis and

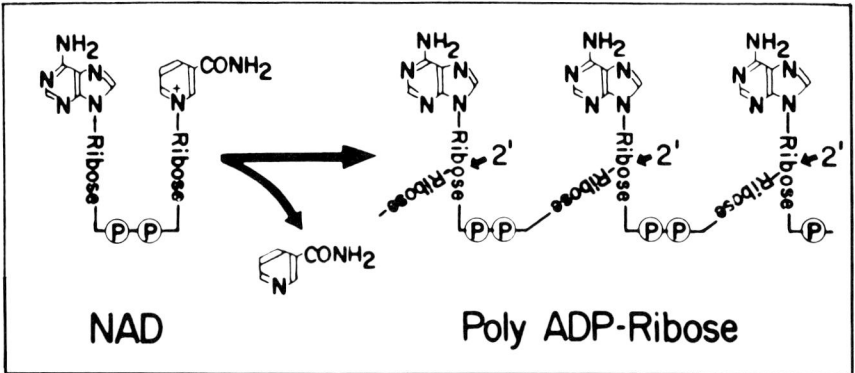

Fig. 10. Synthesis of poly(ADPribose) from NAD with the excision of nicotinamide.

degradation, as well as the polymer itself, are exclusively localized in the nucleus of developing limb bud cells. In order to detect the synthesis of this molecule, intact cells can be incubated with radioactively labeled adenine. Such radioactive adenine will be incorporated internally into cellular NAD and thus can be traced in the biosynthesis of poly(ADPribose). To assay for the cellular synthesis of poly(ADPribose) under these labeling conditions, nuclei are isolated and exposed to RNase and DNase, since poly(ADPribose) is insensitive to these two degradative enzymes. The remaining radioactively labeled nucleic acid can then be exposed to snake venom phosphodiesterase, which specifically cleaves at the phosphate–phosphate diester bridge of poly(ADPribose) and liberates a unique hydrolysis product phosphoribosylAMP.

In experiments pictured in Figure 11, the rate of synthesis of poly(ADPribose) was monitored daily in cultures of chick limb mesenchymal cells [Caplan and Rosenberg 1975a, b; Caplan et al, 1979a, b; Caplan, in press]. As can be seen from Figure 11, the rate of synthesis increased during the early days of culture when these cells were committing to particular phenotypes; then this high rate of synthesis gradually returned to basal level as the cells are in the process of expressing their committed phenotype. When such cultures are exposed to 5-bromodeoxyuridine continuously from day 1 of culture, the rate of poly(ADPribose) synthesis is unchanged and remains at basal levels. It is well documented that exposure to 5-bromodeoxyuridine completely inhibits the commitment event so that these cells cannot enter particular phenotypic pathways [Levitt and Dorfman, 1972; Coleman et al, 1970; Abbott and Holtzer, 1968; Strom and Dorfman, 1976]; however, the cells remain viable and are capable of dividing. Except for synthesis of phenotype-specific macromolecules, these treated cells are relatively healthy. These experimental measurements indicate that there is a correlation between the synthesis of poly(ADPribose) and the commitment of chick limb mesenchymal cells into particular phenotype pathways.

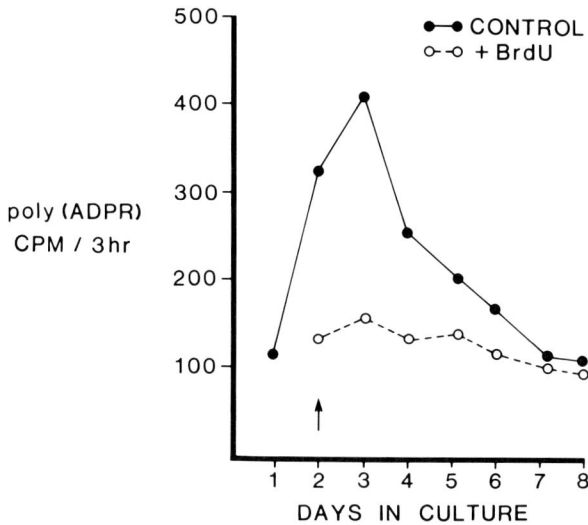

Fig. 11. The rate of synthesis of poly(ADPribose) as a function of days in culture of stage 24 limb mesenchymal cells. ^{3}H-adenine was added to the medium of cultures. After one, three, and five hours, nuclei were isolated and exposed to RNase, DNase and KOH. The cpm sensitive to snake venom phosphodiesterase are plotted per three-hour interval on each day of culture. On day 2, duplicate sister cultures were exposed (arrow) to 10^{-6} M 5-bromodeoxyuridine (BrdU) continuously.

VIII. SUMMARY: THE BIG PICTURE

The information above indicates that chick limb mesenchymal cells can be considered pluripotent for various limb phenotypes. When these cells are exposed to different extracellular environments they respond by differentiating into a particular phenotypic pathway. Of special experimental interest is the fact that the extracellular nicotinamide levels seem to be involved in control of decisional events involved in the commitment of phenotypic pathways of limb mesenchymal cells. Nicotinamide levels structure cellular NAD levels, and both of these informational parameters seem to be related to the synthesis of poly(ADPribose) and the subsequent modification of chromosomal proteins. Thus, an informational transfer system has been identified. This transfer system operates from outside the cell, in terms of nicotinamide levels, to inside the cell, in terms of NAD levels, to interactions with the genetic material and the synthesis of poly(ADPribose), which seems to be correlated with the commitment of limb mesenchymal cells in the particular phenotypes. The physiological basis for differential extracellular nicotinamide levels seems to be the establishment of nutrient flow compartments by the developing vascular system. The avascular core compartment provides informational cues, which are involved in the differentiation of these mesenchymal cells into chondrogenic phenotypes, while the highly vascularized, high nutrient flow compartment in the peripheral tissue is the site for myogenic development.

Based on an elaborately speculative model [Caplan and Ordahl, 1978], I would hypothesize that the following sequence of events occurs in developing limb mesenchyme: a family of pluripotent limb mesenchymal cells has after a series of discrete decisional events, found itself in the growing limb bud. The cells of this family are capable of differentiating into a restricted family of phenotypes; for discussional purposes, we'll consider only muscle and cartilage development. The genes coding for specialized macromolecules indigenous to either the muscle or cartilage phenotype are transcriptively accessible and are sensitive to environmental cues with regard to which genes will predominate. The developing vascular system sets up a differential nutrient flow and provides environmental cues to the developing limb cells. One such environmental cue is the extracellular availability of nicotinamide, which determines the NAD pool size and subsequently has an affect on the synthesis of chromatin associated with poly(ADPribose). This positional information, in addition to other such cues, is interpreted by the cell to cause the repression of one gene set and the augmentation of the specific gene sets that characterize the emerging phenotype. This irreversible repression event marks the commitment phase of development of the phenotypes. The expressional phase follows this commitment event and is governed by other parameters. Thus, based on the limb mesenchymal cell's unique developmental history, its genetic program, and its evolutionary bias, this particular sequence of events leads to the cell's decision to become muscle or cartilage and is followed by the expressional phase of development.

IX. THE EXPRESSIONAL PHASE OF CHONDROGENIC DEVELOPMENT

The high-density cell culture system of stage 24 limb mesenchymal cells provides a superb model from which to study the expressional transitions of developing chondrocytes. Because the majority of cells under these conditions commit to the chondrogenic pathway, we can study the detailed biosynthesis of particular macromolecules during the life cycle of such cells. This can be accomplished by introduction of radioactive tracers on particular days of development and subsequent isolation and purification of the particular macromolecules of interest; the subsequent detailed chemical characterization of these macromolecules can allow us to determine if the biosynthesis of these macromolecules changes in a developmental and time-dependent manner.

With this in mind, Vincent Hascall's laboratory and my laboratory have been involved in the detailed analysis of proteoglycan biosynthesis during the expressional phases of chondrogenic development [Hascall et al, 1976; DeLuca et al, 1977; Kimura et al, 1978; DeLuca et al, 1978; Caplan and Hascall, 1980]. On each day of culture, high-density limb cells were pulsed with ^{35}S-SO$_4$ and ^{3}H-serine and the proteoglycans extracted with high salt and purified via high salt equilibrium density centrifugation with the isolation of 90% of the newly synthesized proteoglycan at the bottom of such gradients. Subsequently, these monomer proteoglycans have been analyzed in detail with regard to their conformation and carbohydrate constituents.

Cartilage proteoglycan is made up of high molecular weight monomeric units held in non-covalent association with an organizing strand of high molecular weight hyaluronic acid [Hascall and Heinegard, 1979; Hascall, 1977]. This relationship is pictured in Figure 12: the monomer proteoglycan unit is made up of a core protein of about 200,000 daltons with chondroitin sulfate and keratan sulfate carbohydrate chains covalently attached to this core protein. We have characterized the monomer proteoglycan from day 8 high-density chick limb bud mesenchymal cell culture, since this day represented the maximum rate of incorporation of ^{35}S-sulfate into proteoglycans. The molecular details of this monomer preparation are pictured in Figure 13. The chondroitin sulfate chains are about 25,000 daltons, the keratan sulfate chains are about 6,000 daltons; and there are approximately 100 chondroitin sulfate chains and about 50 keratan sulfate chains per individual core protein macromolecule. The chondroitin sulfate chains are clustered at one end of the molecule next to a region of clustered keratan sulfate chains, which are next to a protein-rich region that specifically interacts in a non-covalent manner with hyaluronic acid.

Having done the detailed biochemistry of these proteoglycan preparations synthesized on day 8 of culture, we then asked whether proteoglycan synthesized by the newly emerging chondrocyte and the older chondrocytes had different or similar chemical structures. All of these analytical data have been summarized in a series of publications from our laboratories [Hascall et al, 1976; DeLuca et al, 1977; Kimura et al, 1978; DeLuca et al, 1978; Caplan and Hascall, 1980], and the cartoon summary of these experiments is represented in Figure 13. The newly emerging chondrocyte synthesizes proteoglycan with very long chondroitin sulfate chains and very short keratan sulfate chains, and the hyaluronic acid-binding region does not effectively interact with hyaluronic acid. The older chondrocyte, which we refer to as a "senescent chondrocyte," synthesizes a proteoglycan that is hydrodynamically smaller and contains chondroitin sulfate chains that are smaller than the day 8 synthesized material and keratan sulfate chains that are also slightly smaller than those synthesized on day 8. In addition, there has been a change from predominantly 6-sulfated chondroitin to a more equal distribution of 6- and 4-sulfated chondroitin as the cell progresses through development from a newly emerging chondrocyte to a mature or "adult chondrocyte." It should be stressed that our analysis of the isolated and purified

56   Caplan

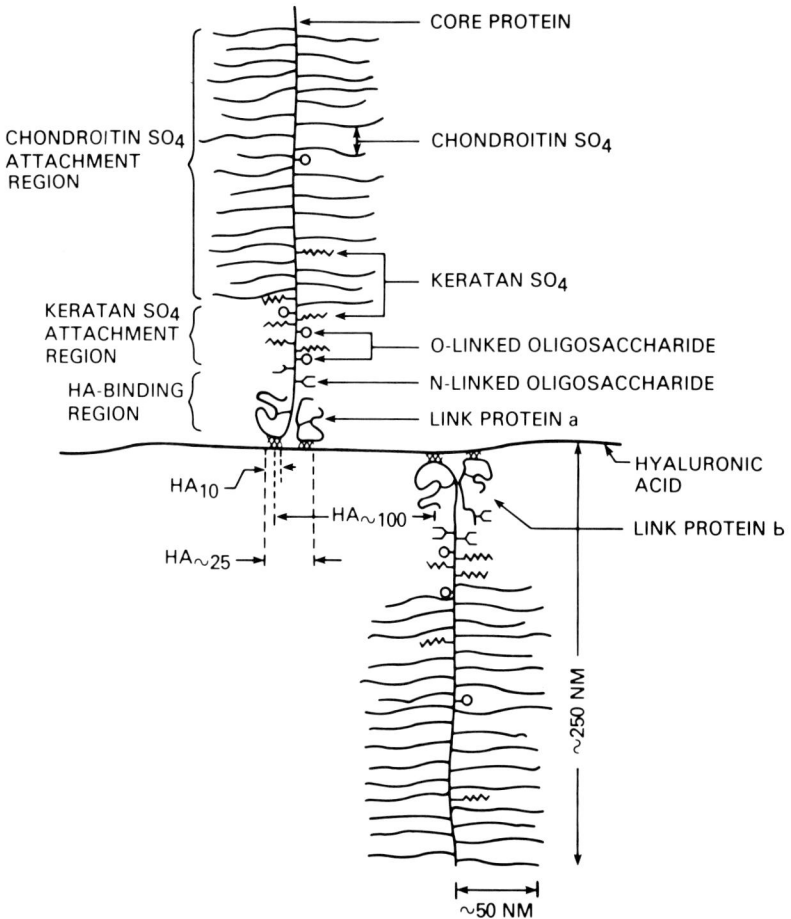

Fig. 12. Proteoglycan monomers as they are situated in aggregate form held together by the non-covalent association of the hyaluronic acid binding region of the proteoglycan core protein and a large strand of hyaluronic acid. The "link" protein stabilizes this noncovalent interaction.

proteoglycan molecules is restricted exclusively to an analysis of the *newly synthesized* molecules on different days of development, since we are characterizing radioactive label incorporated into these molecules on different days of development.

In summary, it is clear that, although type II collagen synthesis is more or less constant during the life history of the chick limb bud chondrocyte culture, the synthesis of the proteoglycan is distinctly different in a progressive manner from the newly emerging chondrocyte through "adulthood" to "senescence."

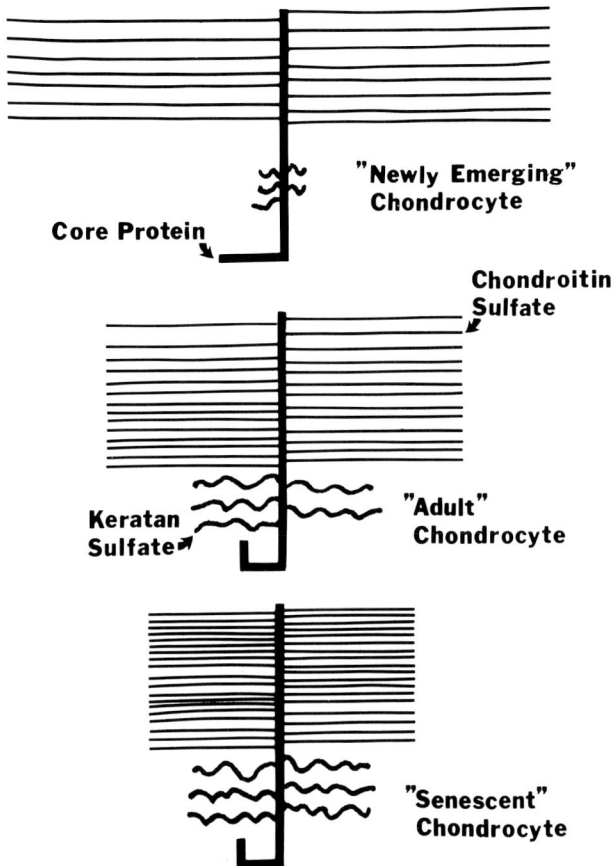

Fig. 13. Schematic representation of cartilage-specific monomer proteoglycan synthesized by day 2 limb bud chondrocytes ("newly emerging" chondrocytes), by day 8 cells ("adult" chondrocytes), and by day 21 cells ("senescent" chondrocytes) in cultures initially seeded at high density with stage 24 limb mesenchymal cells. Data used to formulate this model are published in Hascall et al [1976], DeLuca et al [1977], and Caplan and Hascall [1980]).

Our analysis is precise enough that if we were given a preparation of newly synthesized chick limb chondrocyte proteoglycan, through chemical characterization of the hyaluronic acid binding region, the chondroitin sulfate chains, the keratan sulfate chains, and the monomer size, we could estimate with a high degree of precision the actual day in culture that this newly synthesized proteoglycan was made. If one assumes that the core protein of the proteoglycan is identical during the life history of the chondrocyte, then one must conclude that there is a complex and highly precise regulatory mechanism governing the biosynthesis of cartilage-specific proteoglycan. In addition, since the synthesis of this molecule changes gradually, there must be some mechanism that accounts for this gradual alteration in the actual details of the biosynthesis of these macromolecules.

X. A TEST OF THE HYPOTHESIS: INFLUENCE OF EXTRACELLULAR MATRIX ON GENE EXPRESSION

Armed with the ability to determine analytically the chemistry and structure of newly synthesized proteoglycans, we can use this technology to address a variety of hypotheses generated to explain the developmental transitions experienced during chondrogenic expression. One such theory can be referred to as the "control or influence of the extracellular matrix on gene expression" theory [Slavkin and Gruelich, 1975]. Given the observations of a changing pattern of synthesis of proteoglycans, this theory would suggest that the newly synthesized proteoglycan has a direct feedback effect on the biosynthetic mechanism and that this feedback alters the regulatory steps governing the synthesis of proteoglycan. For example, I would expect that the theory would demand that the first molecule of proteoglycan synthesized by the newly emerging chondrocyte be transported to the matrix and appear in the extracellular space. Its appearance in the extracellular matrix would be sensed by the cells, and the biosynthetic apparatus would be slightly modified as a result of the appearance of the first proteoglycan molecules in this extracellular matrix. The next group of synthesized proteoglycans would be slightly different, and these would then be transported to the matrix. Their appearance in this matrix would again be sensed by the cells, which would respond by again altering the biosynthetic pathway. This continuum of sensing and slight modifications in the biosynthetic pathway could account for the observed changes in the macromolecular structure of newly synthesized proteoglycans.

We have constructed three tests of the theory that the extracellular matrix controls gene expression; in particular, we have constructed a test to ask if extracellular proteoglycan in any way affects the biosynthesis of proteoglycan.

A. Matrix Clearance

We have found conditions in which day 8 high-density cultures can be exposed to a variety of digestive enzymes, including trypsin, chondroitinase ABC, and hyaluronidase, without lifting the cells off of the Petri dish. By labeling such cultures with ^{35}S-sulfate and ^{3}H-serine prior to exposure to these enzymes, we have followed the release of matrix macromolecules during the time course of action of these digestive enzymes [Caplan, DeLuca, and Hascall, unpublished].

At the point where 80–95% of the extracellular proteoglycan and protein is cleared from the matrix, we have terminated exposure to the digestive enzymes and replaced them with normal Complete Medium. After an hour or two of incubation in Complete Medium, the cells are exposed to ^{35}S-sulfate and ^{3}H-serine for three to six hours; the proteoglycans thus synthesized are isolated and purified via cesium chloride equilibrium centrifugation; and the proteoglycan monomer is isolated, purified, and characterized. One could predict two extremes of results from such an experiment: first, one could expect that the day 8 chondrocytes continue to make day 8 proteoglycan in the absence of the normal extracellular matrix, or second, the chondrocytes sense this depletion and the absence of extracellular matrix and because of this, the chondrocytes start the biosynthetic program over. In this latter case, we would observe proteoglycans similar in morphology to those synthesized by the newly emerging chondrocytes. The results of this experiment are summarized later.

B. Replating and Purification of Day 8 Chondrocytes

The second test of the extracellular-matrix-control-of-gene expression hypothesis is to completely remove the "adult" chondrocyte from its extracellular matrix and to place this cell on a naked tissue culture plate. Recently, we have developed a technique whereby, through sequential collagenase digestion of day 8 chick limb mesenchymal cell high-density cultures, we can isolate in pure form either nonchondrogenic cells or pure chondrocytes having less than 3–5% nonchondrocytes [Carrino, Lennon, and Caplan, unpublished]. In this case, the replating of these chondrocytes can be followed by an exposure to ^{35}S-sulfate and ^{3}H-serine to reveal the biosynthesis of proteoglycan from newly attached, mature, chick limb bud chondrocytes. As in the above example, in the extreme there are two possible results: that these replated cells will make day 8 or 9 proteoglycan or that they will restart the program from the beginning and make a proteoglycan similar to that made by the newly emerging chondrocyte. The results of this experiment are summarized later.

C. A Normal Chondrocyte in an Abnormal Extracellular Matrix

When chondrocytes are exposed to 4-methyl umbelliferyl-β-D-xylopyranoside (heretofore referred to as methylxyloside), this compound enters the cell and is recognized by the first galactosyl transferase in the pathway of chondroitin sulfate biosynthesis [Okayamata et al, 1973; Schwartz et al, 1974; Schwartz, 1977; Lohmander et al, 1979a]. A chondrocyte will synthesize a complete chondroitin sulfate chain attached to the methylxyloside, package it, transport it to the external limits of the cell, and secrete it into the extracellular matrix. Because this chondroitin sulfate chain is not attached to the proteoglycan core

END PRODUCT IN

ABSENCE OF   PRESENCE OF

β-D-XYLOSIDE

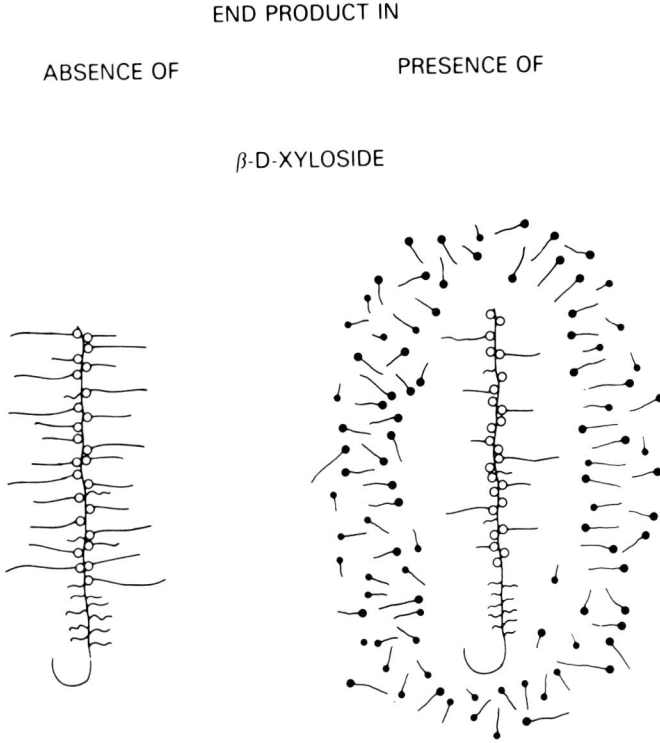

Fig. 14. Model of the proteoglycan synthesized by day 8, high-density cultures of stage 24 limb mesenchymal cells (85% chondrocytes) in the absence and presence of β-D-xyloside. In the presence of xyloside, chondrocytes synthesize a proteoglycan monomer that has the normal complement of keratan sulfate side chains, but the majority of the chondroitin sulfate chains are not associated with the proteoglycan core protein since they are bound to the xyloside (dark circles). These xyloside-associated chondroitin sulfate chains are, thus, found in the tissue culture medium.

protein, it will not integrate with the extracellular matrix. In this case, with cells in culture, such chondroitin sulfate chains bound to the methylxyloside can be found in the medium bathing the cells and can be isolated and characterized [Lohmander et al, 1979b]. Because the methylxyloside *competes* with core protein for the biosynthesis of chondroitin sulfate chains, the proteoglycan monomers synthesized in the presence of methylxyloside are abnormal. Figure 14 summarizes the results of an analysis made of chondrocytes exposed to methylxyloside. Such exposure can be chronic (meaning exposure from the initial plating of uncommitted stage 24 limb mesenchymal cells continuously through day 8) or acute (one hour pretreatment before day 8 cells are exposed to labeled sulfate in the presence of methylxyloside). As pictured in Figure 14, chronically or acutely exposed cells synthesized proteoglycan on day 8 of culture that has approximately two-thirds fewer chondroitin sulfate chains; and those chondroitin sulfate chains that are present are much shorter than normal. Such a decrease in the number of chondroitin sulfate chains was not observed for the number of keratan sulfate chains, which were identical to that of the untreated control. However because there is intense competition for UDP-sugars in the biosynthesis of these macromolecules, the keratan sulfate chains are slightly smaller. Since the keratan sulfate chains are not synthesized in a pathway in which methylxyloside can compete, the major biosynthetic alteration is in the chondroitin sulfate synthesis pathway.

As pictured in Figure 15, the morphology of the extracellular matrix of cells that have been continuously exposed to methylxyloside is strikingly different than that of untreated control cultures. In untreated control cultures, the individual chondrocytes are surrounded by a large proteoglycan- and collagen-filled extracellular matrix. In the chronically treated methylxyloside cultures, the extracellular matrix is almost nonexistent, with individual chondrocytes in close proximity to each other and very little extracellular matrix separating these individual cells. Detailed chemical analysis of such cultures indicates that the amount and type of collagen (ie, type II collagen) [von der Mark and Caplan, unpublished] and the amount of proteoglycan monomer are approximately identical to that of the control cultures. The most obvious conclusion from the morphological observations alone is that the chondroitin sulfate chain attached to the core protein monomer is the primary determinant of the water-structuring quality of the cartilage extracellular matrix. Without these chondroitin sulfate chains, the extracellular matrix structures very little water, and the individual chondrocytes come to lie very close to one another.

Since methylxyloside can be washed out of chondrocytes and the cell can then synthesize proteoglycan in the absence of this competitive factor, one can take chronically exposed cultures on day 8 and terminate the exposure to methylxyloside. After an hour or two of recovery in Complete Medium, the cells were exposed to radioactive tracer, proteoglycan isolated, purified, and characterized in detail. Such treatment would provide a chondrocyte with relatively unaffected biosynthetic pathways but would ask the cell to synthesize cartilage-specific macromolcules while embedded in a totally abnormal extracellular matrix. Just as in the above experiment, one could predict two extremes of response to these environmental conditions: either the chondrocyte would synthesize a normal day 8 proteoglycan or it would sense its abnormal, surrounding matrix and synthesize a proteoglycan somewhere between the newly emerging chondrocyte and the day 8 type proteoglycan.

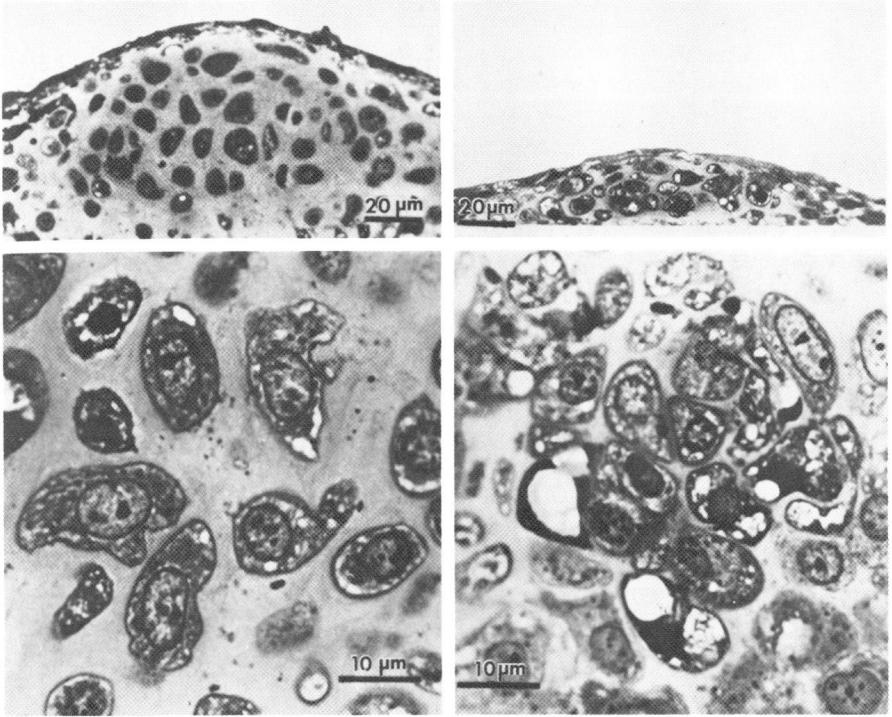

Fig. 15. Comparison between day 8 control (left) and β-D-xyloside-treated (day 1 continuously to day 8), stage 24, limb mesenchymal high-density cell cultures (right). The xyloside-treated cultures have about the same number of cells, but they are situated very close to one another because the cartilage matrix lacks 60–75% of the total synthesized chondroitin sulfate chains found in the culture medium [Lohmander et al, 1979a].

D. Summary of Experimental Results From the Test of the Extracellular Matrix Control of Gene Expression

In all three cases, as detailed above, the matrix-cleared chondrocyte, the replated and purified chondrocyte, and the chondrocyte released from chronic exposure to methylxyloside, *all* synthesize a proteoglycan indistinguishable from normal day 8 proteoglycan. These results all strongly indicate that the expressional changes that we observed for the developing chondrocyte are programed into this cell and that each step along the program represents essentially an irreversible step along the expressional pathway. These three experiments attempt to force the developing chondrocyte to start its biosynthetic program from the beginning; in all three cases and others not reported here, we have not been successful in causing the chick limb bud chondrocyte in culture to reinitiate previous expressional capacities. It is clear that the presence of proteoglycan in the extracellular matrix does not seem to have a controlling influence on the changing biosynthetic program involved in proteoglycan biosynthesis. In addition, I would maintain that this expressional program represents a continuum of steps that are essentially irreversible. It is clear that one can interrupt this biosynthetic program and cause a cell to lose its specialized capacity, but when such biosynthetic capacities are reactivated, the cell essentially picks up where it should be on its expressional program.

One further implication arises from the characterization of the proteoglycan synthesized by chick limb bud cells in culture, related to the structure of the proteoglycan synthesized by the old, or "senescent," chondrocyte. In this case, the monomer size, as indicated by its hydrodynamic radius on molecular sieve columns, is much smaller than that synthesized by the day 8 chondrocyte. Again, this is due primarily to the smaller size of the chondroitin sulfate chain covalently attached to the core protein of proteoglycan. This smaller monomer is, thus, less able to structure or organize water around the highly charged sulfate groups of the chondroitin sulfate chains. This reduced capacity to structure water results in a reduced capacity to provide resiliency of the cartilage matrix. If one were to assume that osteoarthritis, which affects over 95% of humans above the age of 65, is caused by the inability of the joint cartilage to act as a resilient cushion, one could speculate that this develops because the chondrocytes had synthesized "senescent" proteoglycan, which was less able to structure water and, therefore, produces a less resilient cartilage cushion. With a less functional cartilage cushion, injury would result from normal use of joints, inflammation would then occur, and this would be followed by symptoms diagnosed as osteoarthritis.

Assuming that chick limb bud chondrocytes and human chondrocytes have similar expressional programs, can we design useful treatment protocols for patients suffering from osteoarthritis which take this biosynthetic information into account? Some orthopedic surgeons have taken "healthy" cartilage from other joint areas and transplanted it to evacuated, diseased joint areas in older patients. In this case, short-term healing and normal function of these affected joints was observed, but usually this was not a long-term solution. I would suggest that one would need to populate the affected joint with "young" chondrocytes, as opposed to senescent chondrocytes, so that these chondrocytes could produce a proteoglycan that could then structure water and act as a reasonable cushion. One way to do this would be to actually cause a break in the bone, since it is well known that bone repair is initiated with a formation of a cartilage to mend the two pieces of bone together. This newly formed cartilage is embryonic in type and would be exactly that which would be required to re-form a resilient joint cushion. Indeed, some orthopedic surgeons have remarked that when excavating a diseased joint, they have caused small breaks in the tip of the bones and that in these cases, the resulting cartilage cushion that forms seems to be "good as new".

This discussion, which is, of course, highly speculative, is based upon my firm belief that development continues for the entire lifetime of the animal, even past adulthood. It is my belief that there is a genetic program that constitutes the continuum through our embryonic, adolescent, adult, and senescent lifetimes. Based on the observation with chondrocytes, I would suggest that this program is under a tightly regulated temporal program through the entire expressional history of committed and specialized cells. Unfortunately, our ability to survive above the age of 65 is a relatively new phenomenon, and we have not been given an evolutionarily appropriate selection time that allows us to select for people who have the evolutionary advantage of growing old gracefully. Our previous evolutionary history has been predicated on our ability to survive and, indeed, a hundred years ago, which is a relatively short time in evolution, our average life expectancy was half what it is now.

XI. A GENERAL THEME OF DEVELOPMENT

From the above discussion a general theme emerges. That theme is based on a realization that within the genome there is an informational program that governs the entire life history of the organism from fertilization through senescence. This program is marked by distinctive and particular developmental states. Shortly after fertilization early developmental events involve the reorganization of maternal influences and the early expression of the composite male and female genome to produce an embryo of a particular form with no particular phenotypes. One of the early events is the segregation of the germ cell line to ensure the further replication of the species. The events that follow are designed to ensure that the complex functional attributes of the composite organism are protected and these individual functions are segregated into different organ compartments. Cells then become committed to particular developmental pathways and, once committed to a pathway, the cell can no longer regress to this prior decisional state.

The entrance of cells into particular phenotypic compartments is preceded by a long and involved chain of decisions. Early mesenchymal cells actually arise from epithelial cells, and various mesenchymal cells subsequently lose their ability to form a particular organ. For example, although heart muscle has a number of similarities to skeletal muscle, and although they are both of the same mesenchymal origin, the developmental history, and, indeed, some macromolecules synthesized by heart cells and skeletal muscle cells, are uniquely and distinctly different. Likewise when cells become committed to a particular phenotypic pathway and start expressing that commitment, there seems to be a detailed and highly regulated expressional program that has distinct embryonic, adolescent, adult, and senescent stages. These changes are most easily recognized as the cells go from pluripotent cell to a newly emerging phenotype to an adult cell. The changes during the adult cell lifetime and at the entrance of senescence are over such long time spans that it is more difficult to perceive the changes, although we are becoming increasingly more appreciative of them.

It is clear that a number of factors are influential in providing information to the genome which allows it to proceed through these developmental transitions. It is particularly clear that, during chick limb bud mesenchymal commitment, extracellular levels of nicotinamide, packing density, and nutrient flow are important parameters in communicating to the genetic program the physical and temporal position of the cell, and this information is translated into a particular phenotype. It is clear that, during the expressional history of the cell, nicotinamide levels, packing density, and nutrient accessibility are not highly important parameters in the expressional program of the cell. For example, the later stages of muscle development are equally successful at high or low density in the presence of a wide range of nutrient accessibility. The presence or absence of high or low levels of nicotinamide during the expressional phases of muscle development are less important than during the commitment phase.

Although there are factors in the extracellular environment that are highly influential on the commitment phase, it is not quite clear what factors are important in the expressional control of particular phenotypes. Our experimentation with regard to the developing chondrocytes indicates that there does not seem to be a direct feedback effect of proteoglycan in the extracellular matrix on proteoglycan biosynthesis; this is also probably true for collagen. However, the literature now indicates that various effector molecules such as mitogens, growth factors, prostaglandins, and somatomedins are influential on the expressional properties of the individual cells and that these circulating extracellular materials may play an influential or regulatory role in the expressional maturation of individual phenotypes or, at the minimum, on the short-term biosynthetic capacity and functioning of these individual pheontypes.

The realization that the expressional capacities of the cell are under different influences from the developmental capacities of the cell stimulates me to emphasize the difference between a *physiological response* of the cell and a *developmental response*. The differentiated and specialized cell must have the capacity to respond to different physiological stimuli. These stimuli come and go as the general state of the organism changes, and thus physiological reactions must be reversible. One would not want to have a developmental program susceptible to reversible interactions and, therefore, it is my belief that the developmental program is under one-way regulation where cells and processes can proceed only in a forward direction and are not reversible. I think the irreversibility of the developmental phenomenon insures that the organism will be able to reproduce its species without gross involvement of the changing milieu in which this phenomenon takes place. This, however, is not what one would want to build into physiological response of a particular phenotype. In this case, if the nutrient state of the organisms is affected for a relatively short time, one would want to have the individual phenotypes able to respond to these momentary stresses in a reversible way, assuming that the organism will correct this deficiency. This difference between the physiological responsive system and the developmental program is highly important and distinguishes between how eukaryotic organisms have evolved successful developmental programs. This is distinctly different from the physiologically responsive program found in E coli, which does not have a developmental program. When specialized functions are required, it seems that one would need E. coli type cells that are capable of reproduction of that committed phenotype and that are physiologically responsive. In forming the different families of physiologically responsive cells, one would like to have that developmental program under tight and irreversible, one-way regulation to ensure that the particular segregation of specialization takes place in a finite period and under strict genetic control. Again, the evolutionary histories of eukaryotic cells have played a substantial role in structuring the genetic program to include two separable programs, developmental and physiological, which are regulated by different molecular mechanisms. The integration of these genetic programs with the extracellular environment forms the basis for a successful developmental progression of a physiological responsive organism.

ACKNOWLEDGMENTS

I wish to express my appreciation to the Society of Developmental Biology for inviting me to participate. This manuscript is dedicated to honor Walter Landauer, whose 1957 paper opened the door to experimental embryology for me. The force of his deductive logic and the elegant and uncluttered experimentation made the general implications of his scholarly efforts obvious. I would also like to honor Edgar Zwilling, the soft and gentle man in whose laboratory this early work was done. I still feel orphaned by his passing ten years ago.

Some of the work reported here was supported by grants from N.I.H., March of Dimes Birth Defects Foundation, and the Muscular Dystrophy Association of America.

REFERENCES

Abbott J, Holtzer H (1968). Proc Natl Acad Sci USA 59:1144–1151.

Caplan AI (1970). Exp Cell Res 62:341–355.

Caplan AI (1972). J Exp Zool 180:351–362.

Caplan AI (in press). In "Poly (ADPRibose) and ADPribosylated Proteins." Monograph. Elsevier, North Holland.

Caplan AI, Hascall VC (1980). In "Dilation of the Uterine Cervix." (F Naftolin and Stubblefield, eds.), Raven Press, New York. pp 79–98.

Caplan AI, Koutroupas S (1973). J Embryol Exp Morphol 29:571–583.

Caplan AI, Ordahl CP (1978). Science 201:120–130.

Caplan AI, Rosenberg M (1975a). In "Extracellular Matrix Influences on Gene Expression." (H. Slavkin and R Gruelich, eds), pp. 47–54. Academic Press, New York.

Caplan AI, Rosenberg M (1975b). Proc Natl Acad Sci USA 72:1852–1857.

Caplan AI, Stoolmiller AC (1973). Proc Natl Acad Sci USA 70:1713–1717.

Caplan AI, Zwilling E, Kaplan NO (1968). Science 160:1009–1010.

Caplan AI, Neidergang C, Okazaki H, Mandel P (1979a). Dev Biol 72:102–109.

Caplan AI, Neidergang C, Okazaki H, Mandel P (1979b). Arch Biochem Biophys 198:60–69.

Chambon P, Weill JD, Mandel P (1963). Biochem Biophys Res Commun 11:39–43.

Christ B, Jacob JJ, Jacob M (1977). Anat Embryol 150:171.

Coleman AW, Coleman JR, Kankel D, Werner I (1970). Exp Cell Res 59:319–328.

DeLuca S, Heinegard D, Hascall VC, Kimura J, Caplan AI (1977). J Biol Chem 252:6600–6608.

DeLuca S, Caplan AI, Hascall VC (1978). J Biol Chem 253:4713–4720.

Finch RA, Zwilling E (1971). J Exp Zool 176:397–408.

Hamburger V, Hamilton HL (1951). J Morphol 88:49–92.

Handler P, Klein JR (1942a). J Biol Chem 143:49–57.

Handler P, Klein JR (1942b). J Biol Chem 144:453–454.

Harris M (ed) (1973). In "Poly(ADPR-ibose) An International Symposium" Fogarty Int Cent Proc No 26. Washington, DC: U.S. Government Printing Office.

Hascall VC (1977). J Supramol Struct 7:101–120.

Hascall VC, Heinegard D (1979). In "Glycoconjugate Research: Proc. 4th Int Symp. on Glycoconjugate." (I. Gregory and R Jenaloz, eds.), pp 341–374. Academic Press, New York.

Hascall VC, Oegema TR, Brown M, Caplan AI (1976). J Biol Chem 251:3511–3519.

Hauschka S, Konigsberg IR (1966). Proc Natl Acad Sci USA 55:119–126.

Kaplan NO, Goldin H, Humphreys SR, Ciotti MM, Venditti JM (1954). J Biol Chem 219:287–298.

Kieny M, Chevallier A (1979). J Embryol Exp Morphol 49:153–165.

Kimura JH, Osdoby P, Caplan AI, Hascall VC (1978). J Biol Chem 253:4721–4729.

Konigsberg IR (1960). Exp Cell Res 21:414–420.

Landauer W (1948). J Exp Zool 109:283–290.

Landauer W (1954). J Cell Comp Physiol 43(Suppl 1):261–305.

Landauer W (1957). J Exp Zool 136:509–530.

Landauer W, Clark EM (1962). J Exp Zool 151:253–258.

Landauer W, Wakasugi N (1967). J Exp Zool 164:499–516.

Landauer W, Sopher D (1970). J Embryol Exp Morphol 24:187–202.

Levitt D, Dorfman A (1972). Proc Natl Acad Sci USA 69:1253–1257.

Lohmander S, Hascall VC, Caplan AI (1979a). J Biol Chem 254:10551–10561.

Lohmander S, Madsen K, Hinek A (1979b). Arch Biochem Biophys 192:148–157.

McMahon D (1974). Science 185:1012–1021.

Oide H (1958). Gann 49:49–56.

Okayama H, Edson CM, Fukushima M, Hayaishi O (1976). In "Proceeding of the 4th International Symposium on Poly ADP-Ribose and ADP-ribosylation of Proteins." p 1. De Gruyter, Berlin.

Okayamata M, Kimata K, Suzuki S (1973). J Biochem (Tokyo) 74:1069–1073.

Osdoby P, Caplan AI (1979). Dev Biol 73:84–102.

Osdoby P, Caplan AI (1980). Calcif Tissue Int 30:43–50.

Rosenberg MJ, Caplan AI (1974). Dev Biol 38:157–164.

Rosenberg MJ, Caplan AI (1975). J Embryol Exp Morphol 33:947–956.

Rudnick D (1945). Trans Connecticut Acad Arts Sci 36:353–377.

Saunders J (1948). J Exp Zool 108:363–404.

Schwartz NB (1977). J Biol Chem 252:6316–6321.

Schwartz NB, Galligani L, Ho PL, Dorfman A (1974). Proc Natl Acad Sci USA 71:4047–4051.

Searls R, Janners M (1969). J Exp Zool 170:365–376.

Slavkin H, Gruelich R (1975). Eds. of "Extracellular Matrix Influences on Gene Expression." Academic Press, New York.

Strom CM, Dorfman A (1976). Proc Natl Acad Sci USA 73:1019–1023.

Wolff, E (1936). Arch Anat Histol Embryol 22:1–382.

Zwilling E (1948). J Exp Zool 109:197–214.

Zwilling E (1949). Proc Soc Exp Biol Med 71:609–612.

Zwilling E (1968). Dev Biol Suppl 2:184–207.

Zwilling E DeBell JT (1950). J Exp Zool 115:59–81.

Structure and Organization of Developmentally Regulated Chorion Genes From Antheraea polyphemus

C. Weldon Jones and Fotis C. Kafatos

Cellular and Developmental Biology, Biological Laboratories, Harvard University, Cambridge, Massachusetts 02138

I. INTRODUCTION

The development of a eukaryotic organism is a complex process that depends upon the differential expression of genes over time and space. Each gene involved in a developmental pathway must be strictly regulated so that it can be expressed at the proper time, in coordination with other genes. An understanding of the mechanisms involved in the regulation of eukaryotic gene expression is thus an important component of understanding development. With the advent of recombinant DNA techniques, entirely new approaches toward this goal have become possible. A number of favorable model systems have been used, including the insect chorion.

During the terminal growth phase of the developing insect egg, the chorion, or eggshell, is synthesized by a layer of follicular cells surrounding the oocyte. In silkmoths, the chorion forms a hard, protective layer and consists of over 100 different proteins, as determined by two-dimensional gel electrophoresis. Over the two-day period of choriogenesis, each protein is synthesized during

Levels of Genetic Control in Development, pages 69–81

a characteristic and limited time period, both in vivo and in organ culture. Thus, the ultimate expression of the chorion genes obeys a strict developmental program.

In the silkmoth, Antheraea polyphemus, the chorion proteins may be divided into at least four molecular weight classes: A (7,000–11,000 daltons), B (11,-000–15,000 daltons), C (15,000–20,000 daltons), and D (20,000–30,000 daltons). The bulk of chorion protein synthesis involves production of the A and B proteins, which make up approximately 85% of the eggshell by weight. The C proteins are synthesized primarily during the early stages, while the several A and B proteins are synthesized at various periods throughout choriogenesis.

At least 18 different A proteins and 25 different B proteins have been resolved by two-dimensional gel electrophoresis [Regier et al, 1980]. Several members of these two classes of chorion proteins have been sequenced partially [Regier et al, 1978a,b; Rodakis, 1978]. The A proteins were sequenced from the NH_2-terminus; since the B proteins have a blocked NH_2-terminus, it was necessary to sequence them from an internal methionine residue after cyanogen bromide cleavage. The protein sequencing results revealed that members of each size class are very similar in amino acid sequence, but they do differ from one another by internal amino acid substitutions and small insertions or deletions. The intraclass sequence similarities are so extensive that it is clear that the members of each class are homologous—ie, evolved from a common ancestor by gene duplication followed by diversification. Considerable differences exist between A and B proteins. Therefore, it appears that the A and B proteins are encoded by two developmentally regulated multigene families [Hood et al, 1975].

II. CHORION cDNA CLONES

During choriogenesis most (> 95%) of the mRNAs in the follicular cells encode chorion proteins. This, coupled with their small size, has made it relatively easy to isolate chorion mRNAs as a class. From total chorion mRNA, a plasmid library of double-stranded cDNA clones was constructed, and several clones were characterized in detail [Sim et al, 1979; Jones et al, 1979]. Cross-hybridization studies and direct sequencing of cDNA clone inserts confirmed that chorion mRNAs within each of the major families (A and B) are homologous.

By dot hybridizations, we were able to characterize 22 distinct cDNA clones in terms of a number of different parameters: developmental specificity (ie, time when the corresponding sequence is represented in mRNA), evolutionary history (ie, degree of cross-hybridization with other cDNA clones), and quantitative specificity (ie, abundance of the corresponding mRNA at all developmental stages combined) [Kafatos et al, 1979; Sim et al, 1979]. These studies made the cDNA clones invaluable as probes for studying the structure and organization of chromosomal chorion genes with defined developmental and evolutionary properties (see below).

The nucleic acid studies have extended our knowledge concerning the structure and evolution of the chorion mRNAs and their corresponding proteins. The chorion-coding region of seven A and seven B sequences have been compared [Jones et al, 1979; Tsitilou et al, 1980; Jones, 1980]. The nucleotide (or amino acid) sequences of members of the same family can easily be aligned for much of their length. However, the aligned sequences may be interrupted by small deletions or insertions, often involved in the expansion or contraction of internally repetitive regions. Within a family, the central portion of the coding region is the most highly conserved, with few, if any, insertions or deletions. Sequence regions that are distinct in amino acid composition, as well as in the extent of divergence, have been defined and, presumably, correspond to distinct functional domains [Jones et al, 1979].

The extensive sequence information that has accumulated thanks to cDNA cloning and rapid sequencing techniques reveals the existence of similarities between the two families at the amino acid level. The conserved central portion of the proteins is similar in composition and shows some very limited sequence similarities between families. The extreme COOH-terminal region of both families contains tandem Cys·Gly repeats. Moreover, a region near the COOH-terminus of the B proteins shows sequence similarities, with two regions of the A proteins, one at or near the NH_2-terminus and the other near the COOH-terminus. Since the nucleotide sequence similarities between the two families are limited, it is not possible to state whether the similarities at the amino acid level are the result of convergent or divergent evolution.

III. THE STRUCTURE OF CHORION GENES

A number of clustered multigene families, as well as a number of dispersed multigene families, have been discovered in eukaryotes in recent years. By genetic analysis of the commercial silkmoth, Bombyx mori, the physical linkage of chorion genes in this organism has been established. The chorion genes occur as three neighboring clusters, located within a few map units at one end of a single chromosome (n = 28) [Goldsmith and Clermont-Rattner, 1979].

The resolution of genetic analysis is necessarily low in an organism with a long life cycle such as Bombyx. In the wild silkmoth, A polyphemus, no genetic information is available. However, it has been possible to examine the structure and organization of the chorion genes by recombinant DNA techniques.

To achieve this, a library of cloned A polyphemus chromosomal DNA segments was constructed using the charon 4 derivative of phage λ as vector

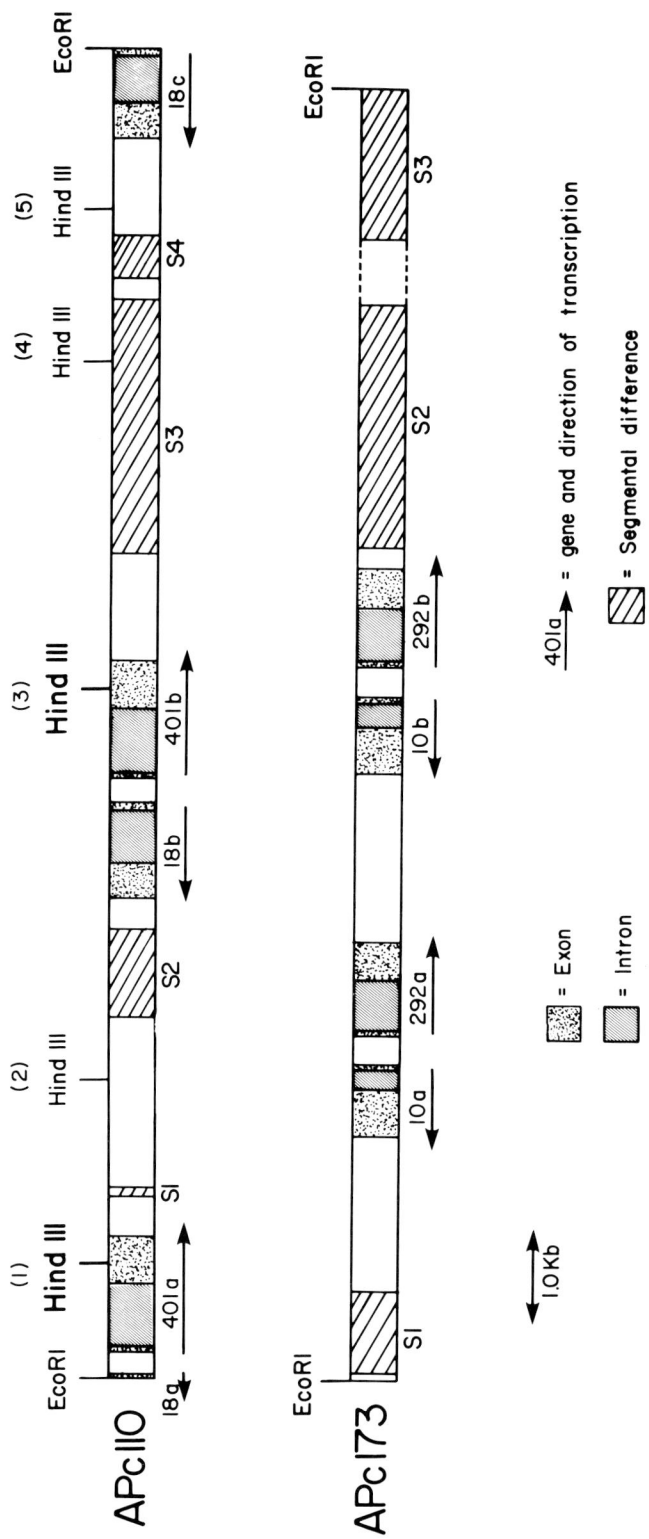

Figure 1.

[Maniatis et al, 1978; Jones and Kafatos, 1980]. The library was screened initially for clones containing chorion genes. Of 175 purified recombinant phages, two cloned segments were selected for further analysis because they each hybridized to two characterized cDNA clones. The cloned chromosomal segments (14 kb each) were designated APc110 and APc173. APc110 hybridizes to the cDNA clone pc401, which is a member of the B family, and to clone pc18, which is a member of the A family. Both of these genes are expressed coordinately: mRNAs for pc401 and pc18 are detected in the cytoplasm in parallel, primarily during the late period of choriogenesis. APc173 hybridizes to cDNA clone pc10 (a B family member), and cDNA clone pc292 (an A). Their mRNAs appear coordinately in the cytoplasm, primarily during the middle period of choriogenesis [Sim et al, 1979; Tsitilou et al, 1980]. Thus, it appears that chorion genes are not segregated according to multigene family, but are clustered according to the developmental period during which they are expressed.

Schematic diagrams of the two cloned genomic segments are shown in Figure 1. APc110 contains two copies of gene 401 (401a and 401b), plus one complete and two partial copies of gene 18 (18b; 18a and 18c, respectively). APc173 contains two complete copies of both genes 10 and 292. The genes belonging to the two multigene families alternate and are divergently transcribed from opposite strands.

Sequence analysis reveals that each gene consists of two exons and a single intron (Fig. 1), which is invariably located between the codons for amino acids -4 and -5 of the signal peptide (counting from the mature protein NH_2-terminus).

The gene copies found on each segment are not identical, differing by base substitutions and small insertions or deletions (Fig. 2). The inter-copy divergence for genes 401 and 18 is only 0.9%, suggesting that the copies probably were duplicated relatively recently. In contrast, the gene 10 and gene 292 copies are more divergent, differing by about 5.5%, suggesting a more ancient duplication event. Nearly all insertion/deletion events that have occurred between gene copies appear to have involved tandem or non-tandem direct repeats.

The introns of gene copies are very similar both in sequence and in length. However, no significant sequence similarity exists for introns of different genes,

---

Fig. 1. Schematic maps of the cloned chromosomal segments APc110 and APc173. Both cloned segments are approximately 14 kb in length. The chorion genes and their direction of transcription are indicated for each. APc110 contains two complete copies of gene 401 (401a and 401b) and one complete and two partial copies of gene 18 (18b, 18a, and 18c, respectively). The organization of APc173 is similar, with two complete copies of both gene 10 (10a and 10b) and gene 292 (292a and 292b). The positions of the exons and intron for each gene, as determined by DNA sequencing, are shown. The only part of 18a contained in APc110 is included in the small exon; 18c is missing this part. Hind III sites (numbered consecutively) located with APc110 are included, which also mark the boundaries of fragments subcloned into the unique Hind III site of pBR322. Use of subcloned Hind III fragments for heteroduplex analysis revealed segmental differences between the homologous 3' flanking regions within both APc110 and APc173 (an example is shown in Fig. 4). The non-homologous segmental differences are indicated by cross-hatching and numbered (S1 to S4).

APc IIO

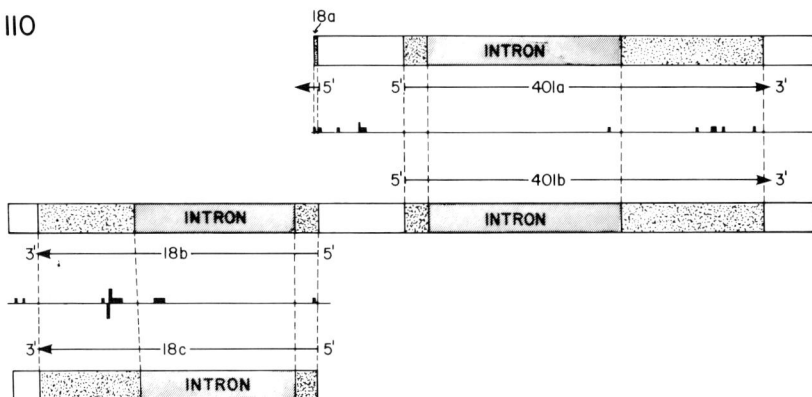

APc I73

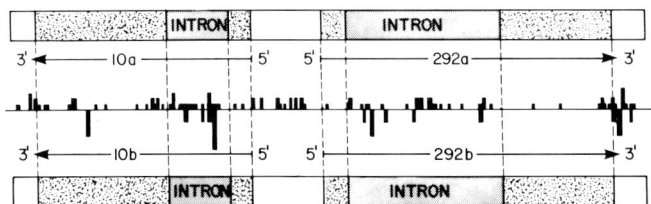

0.25 Kb

Fig. 2. Sequence comparison of segments bearing chorion genes from APc110 and APc173. The positions on APc110 and APc173 of the segments compared here are shown in Figure 1. Each segment was sequenced in its entirety [C.W. Jones, unpublished results]. Between the segments compared are histograms indicating the positions and numbers of base substitutions (above line) and insertions or deletions (below line) between the two homologous segments. Each of the segments has been divided into ten-nucleotide intervals, and any base substitution and/or insertion or deletion that occurs within the interval is indicated. Height is proportional to the number of individual mutational events. Each insertion or deletion event has been arbitrarily given the value of 2.5 base substitution events.

whether belonging to the same multigene family (eg, 10 and 401) or to the same developmental class (eg, 401 and 18). The introns vary in length from 703 bp for 401a and 401b to 232 bp for 10a. Despite their lack of homology, all chorion gene introns are A + T rich, approximately 73%. In agreement with the consensus sequence of sequenced introns (other than those of tRNA genes), the chorion gene introns begin with the sequence *GTPuAG* and end with *AG* [Breathnach et al, 1978]. If these junction sequences are involved in the splicing out of the intron sequence during RNA processing, it would appear that this process is not developmentally stage specific.

The positions of the 5′ ends of each gene were determined by comparison with sequences of full-length cDNAs. For all eight genes and gene copies represented in APc110 and APc173, the sequence at the mRNA cap site (start of transcription?) is PuTCATT. This common sequence is similar to that determined for sea urchin histone mRNAs [Sures et al, 1980]. Again, it would appear that this sequence is a signal for a ubiquitous process such as initiation of transcription or RNA capping and RNA maturation, rather than a developmental stage-specific signal.

The distance between the 5′ ends of the genes belonging to a divergently transcribed gene pair is short, only 325 bp for 401/18 and 264 bp for 10/292. The distance is invariant for different copies. These two facts suggest that the short 5′ flanking regions may be important for the proper expression of the genes.

The portion of the 5′ flanking region immediately upstream from the position of the mRNA cap site is shown in Figure 3. Located 21 to 23 nucleotides upstream from the 5′ end of each gene is the sequence TATA$_A^T$AA (Hogness box) [Gannon et al, 1979]. This sequence, analogous to the prokaryotic Pribnow box, has been found at the corresponding position for all other sequenced eukaryotic protein-coding genes and is thought to be a nonspecific component of RNA polymerase II promotors [Gannon et al, 1979]. Except for the Hogness box, sequence homologies in the 5′ flanking regions are quite limited, both between genes of the same multigene family and between genes expressed during the same developmental period. However, some features of the 5′ flanking sequences may be consistent and presumably functionally important. The 100 nucleotides upstream from each cap site can be subdivided into several short segments (close to ten or 20 nucleotides in length) according to base composition. In general, A + T-rich segments alternate with relatively A-poor segments. The Hogness box is an extremely A + T-rich segment flanked by relatively A + T-poor segments. Beginning at position -76 to -80 from the 5′ end of the genes is a 19-nucleotide region that has some features common to the "-35 recognition sequence" in the prokaryotic promotor region (Fig. 3). The proximal part of this sequence is very A + T-rich, whereas the distal part is an A-poor segment eight nucleotides in length. The latter segment, occurring in a "valley" of A + T content, shows developmental specificity: in the middle genes, 10 and 292, it includes the sequence TG$_A^T$TA, and in the late genes, 401 and 18, it includes the sequence T$_T^A$CGTG. Whether these common sequences have a role in the stage-specific expression of the chorion genes must await the analysis of additional middle and late genes, and ultimately the results of cell-free transcription studies using intact and "engineered" clones.

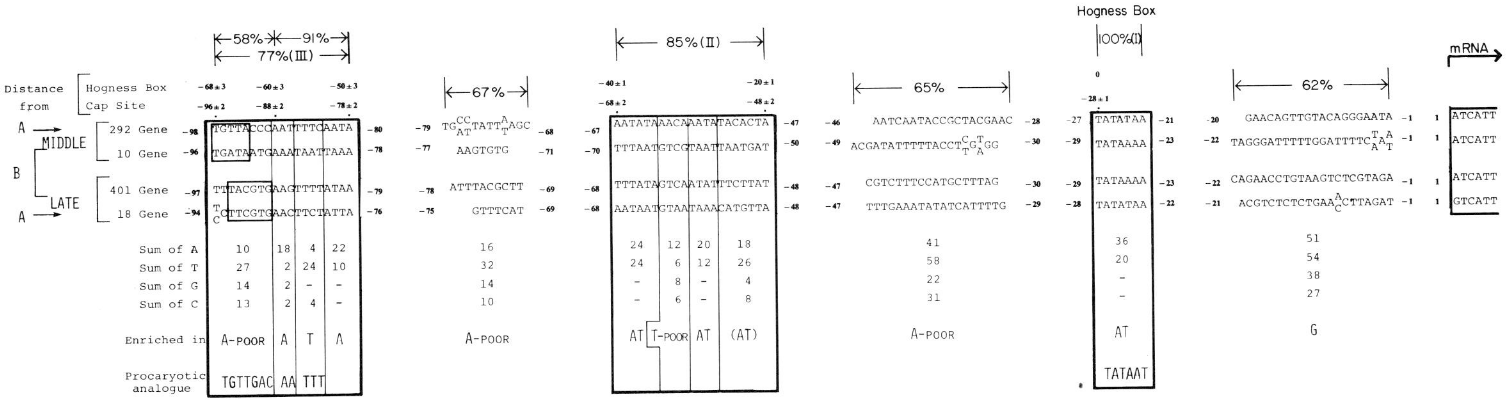

Fig 3. Sequence comparison of 5′ flanking regions of chorion genes. The 5′ flanking sequences of the 401/18 and 10/292 gene pairs are presented. Because of the divergent polarity of the gene pairs, sequences adjacent to the genes of each pair are read from opposite strands, beginning at the respective mRNA cap site (to the right) and extending toward the middle of the 325 bp or 264 bp segment separating the genes (to the left). The sequences of gene copies (eg, 401a and 401b) are identical, except at the position indicated (with both bases shown). The sequences of genes 10 and 292, expressed during the middle period of choriogenesis, are shown above the sequences of genes 401 and 18, expressed during the late period. The sequences have been divided into several small segments according to base composition. The A + T-rich segments are designated I, II, or III and are boxed with thick lines. The percent A + T is given above each of the segments. The sum of each of the four bases for each segment, or sub-segment, is given below the sequences (the numbers include both gene copies). The developmentally specific sequences in region III (see text), TG$_A$TA and T$_A^A$CGTG, are boxed. The boldface numbers refer to the distances in nucleotides from the cap site and/or Hogness box.

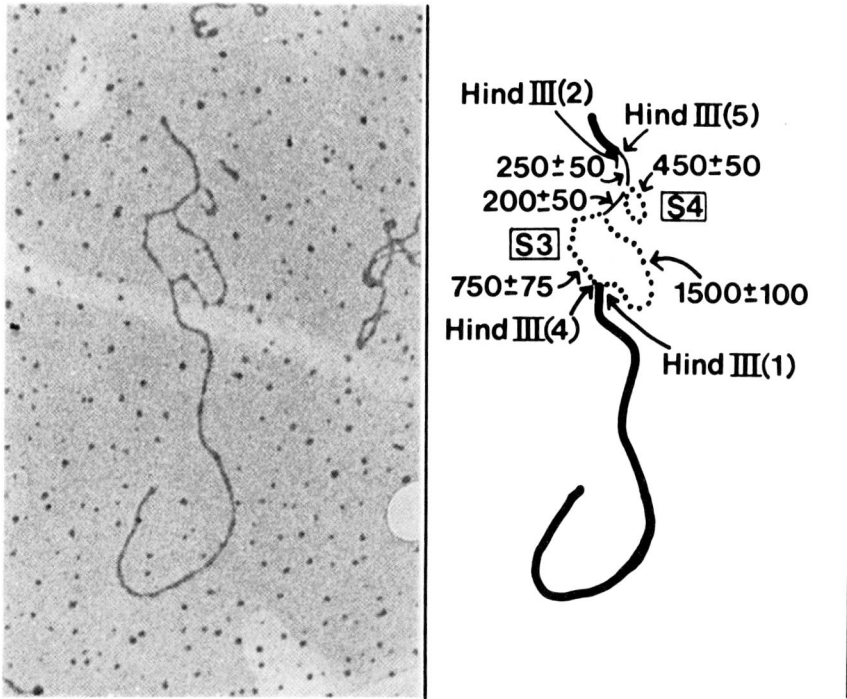

Fig. 4. Electron micrograph of heteroduplex formed between two plasmids containing APc110 Hind III fragments. The positions of the APc110 Hind III fragments subcloned into pBR322 are shown in Figure 1. The two plasmids shown here contain the APc110 fragments extending from Hind III (1) to Hind III (2) and from Hind III (4) to Hind III (5). The plasmids were linearized with the restriction enzyme Sal I, which cleaves pBR322 once, but not the moth-containing insert. The two plasmids were mixed, hybridized, and prepared for electron microscopy. The resulting heteroduplex structure (to the left) is shown with an interpretive sketch (to the right). In the interpretive sketch, pBR322 sequences are shown with thick, solid lines. Double- and single-stranded insert sequences are shown with thin solid lines and dotted lines, respectively. The Hind III sites (pBR322/moth DNA junction) are indicated. The lengths, with standard deviation, for each of the double- and single-stranded segments are presented. The small single-stranded loop is the segmental difference S4, located between Hind III (4) and Hind III (5). A portion of segmental difference S3 is also indicated between Hind III (4) and (5). Much of the 1,500 ± 100 single-stranded segment from the Hind III (1) and (2) fragment is homologous to other subcloned fragments from APc110 [C.W. Jones, unpublished results] (see also Fig. 1).

The extensive sequence homology between gene copies extends beyond the 3′ ends of the genes, according to the results of heteroduplex analysis (for an example, see Fig. 4). The difference in length between the 3′ flanking sequences, separating copies of the gene pairs, is due to the presence of relatively large insertions or deletions, ranging in length from 0.065 kb to 2.9 kb (Fig. 1). In other words, the fundamental unit of chorion gene organization consists of a long tandem repeat, within which one gene pair is embedded. The acceptance of large insertions/deletions in the 3' flanking portion of the repeat (and, according to preliminary data, in the intron regions as well) may be an important aspect of the evolutionary divergence of the chorion genes.

IV. ORGANIZATION OF THE 401/18 GENE PAIR REPEATS

Southern [1975] blot analysis has revealed that a total of ten to 20 copies of the 401/18 gene pairs exist per haploid genome (Fig. 5). Both indirect and direct evidence suggests that these copies of the gene pairs are tandemly organized. Several clones containing 401/18 gene pairs were isolated from a chromosomal library constructed from the DNA of a single moth. Four clones, containing

---

Fig. 5. Detection and organization of repeated 401 and 18 genes in the moth genome. Total DNA from pooled developing pupae was digested with the restriction enzyme Kpn I, electrophoresed, transferred to nitrocellulose paper, and hybridized at a high criterion with 5'- or 3'-specific radio-labeled probes consisting of Kpn I fragments from nick translated cDNA clones pc401 and pc18 (Kpn I cleaves once in the cDNA inserts). Each probe contained the appropriate half of the chorion cDNA insert plus flanking plasmid sequences. Two aliquots of moth DNA were used in each experiment $(1 \times, 10 \mu g/2 \times, 20 \mu g)$. Reconstruction standards were included to estimate the gene multiplicity. They consisted of increasing amounts of a Hind III/Kpn I digest of a subcloned Hind III fragment from APc110 (containing the entire 18b gene and the 5' half of the 401b gene; see Fig. 1), plus 10 μg sheared calf thymus DNA; only the hybridizing Hind III to Kpn I moth DNA fragments are shown. Each standard corresponds to the indicated number of gene copies per haploid genome $(C = 1 \text{ pg})$, relative to the $1 \times$ moth DNA sample.

Two major genomic bands hybridize with both 5' probes (top). When a mixture of the two probes is used, again only two bands are seen, confirming that the 5' ends of both types of genes hybridize to the same fragments (data not shown). The lower band (2.1 kb; dot) is expected from the map of APc110 (Kpn I fragment containing the 5' ends of 18b and 401b; Fig. 1). The upper band (2.5 kb) indicates the existence of a second type of linked and divergent 401/18 pair in A polyphemus. The combined intensities of the two bands, in comparison with the standards, indicate the existence of approximately 15 ± 5 401/18 gene pairs per haploid genome (the gene 18 estimate is more dependable in this respect, since probe, standard, and genomic fragment are coterminous, whereas the standard lacks 130 of the 300 pb of 401 sequence represented in both probe and genomic fragment). At very long exposures, faint hybridization to additional high molecular weight bands can be seen with both probes, possibly indicating limited polymorphism, single copies of additional 401/18 pairs, or cross-hybridization with different but similarly organized genes of the same multigene families [Sim et al, 1979].

Hybridizations with 3'-specific probes (bottom) reveal multiple bands, only some of which are identical for the two genes. This indicates that heterogeneity beyond the 3' ends of the genes is much greater than between the 5' ends. The expected 3.9 kb fragment (between 401a and 18b; Fig. 1) is indicated by a dot; the expected 5.8 kb fragment (between 401b and 18c) is seen in long exposures, but it is smeared because of localized overloading of the gel (it comigrates with a large number of unrelated Kpn I genomic DNA fragments).

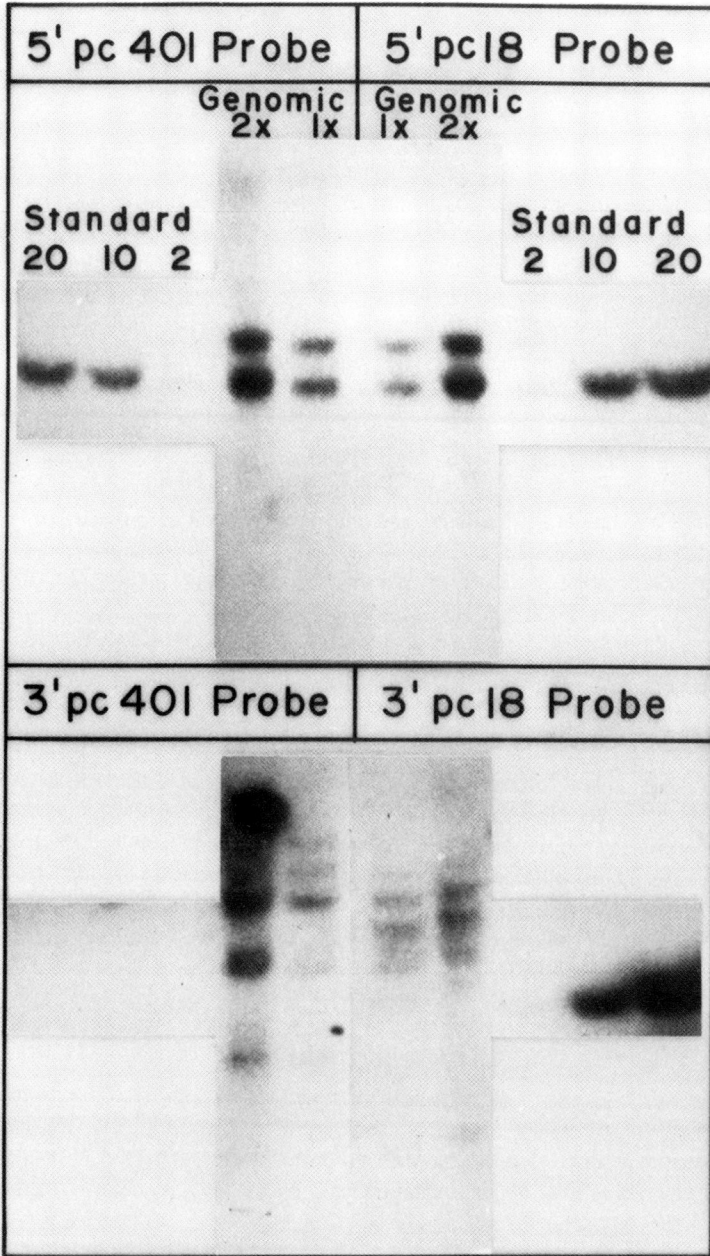

Figure 5.

overlapping fragments and spanning 19 kb, possessed a total of three complete and one partial 401/18 gene pairs [Jones, 1980]. Several additional cloned segments have been shown to contain at least two gene pairs. Because no other chorion genes have been determined to be adjacent to a 401/18 gene pair, it is likely that the ten to 20 copies of these genes are tandemly repeated.

It has been proposed that tandemly repeated genes and their flanking DNA remain homologous by repeated cycles of unequal crossing-over [Smith, 1973]. Present evidence indicates that the non-coding regions of the 401/18 gene pairs are indeed homologous. Southern [1975] filter hybridizations using intron-specific and 5' flanking-specific DNA fragments as probes have shown that copies of the gene introns and 5' flanking DNAs are highly homologous. In addition, Hind III and Sal I restriction enzyme sites located at the same relative positions in the DNA between 401/18 gene pairs suggest that the 3' flanking regions are also homologous. Because the analysis of unlinked chorion genes has indicated that the non-coding regions are generally under much less stringent selective pressure to remain the same than the coding region, it is likely that the 401/18 gene pairs are tandemly repeated and that the observed sequence similarity is due to unequal crossing-over.

Unlike the histone and rRNA genes, the chorion genes do not appear to have short, highly repeated, tandem simple repeats in their 5' or 3' flanking regions [Kedes, 1979; Wellauer et al, 1976]. It has been proposed that the simple repeats between the histone and rRNA genes would aid in allowing a greater frequency of in-register unequal crossing-over events. This would assure homogeneity of gene sequence, which is important for these genes because of the biological significance of their products. In contrast, the 3' flanking regions of the chorion genes are homologous but are interrupted by large insertions or deletions of variable size and sequence. These segmental differences should effectively decrease the rate of unequal crossing-over and result in a more rapid diversification of the chorion gene families. This type of diversification would be acceptable and even beneficial to the silkmoth because of the possibilities it presents for rapid phenotypic evolution.

ACKNOWLEDGMENTS

We thank S.A. Byers for expert secretarial assistance and B. Klumpar for help with the figures. This work was supported by grants from NSF and NIH to F.C. Kafatos. C.W. Jones was supported by an NSF predoctoral fellowship and NIH training grant.

REFERENCES

Breathnach R, Benoist C, O'Hare K, Gannon F, Chambon P (1978). Proc Natl Acad Sci USA 75:4853–4857.

Gannon F, O'Hare K, Perrin F, LePennec JP, Benoist C, Cochet M, Breathnach R, Royal A, Garapin A, Cami B, Chambon P (1979). Nature 278:428–434.

Goldsmith M, Clermont-Rattner E (1979). Genetics 92:1173–1185.

Hood L, Campbell JW, Elgin SCR (1975). Annu Rev Genet 9:305–353.

Jones CW (1980). Ph.D. Thesis, Harvard University, Cambridge, Massachusetts.

Jones CW, Kafatos FC (1980). Nature 284:635–637.

Jones CW, Rosenthal N, Rodakis GC, Kafatos FC (1979). Cell 18:1317–1332.

Kafatos FC, Jones CW, Efstratiadis A (1979). Nucleic Acids Res 7:1541–1552.

Kedes LH (1979). Annu Rev Biochem 48:837–870.

Maniatis T, Hardison RC, Lacy E, Lauer J, O'Connell C, Quon D, Sim G-K, Efstratiadis A (1978). Cell 15:687–701.

Regier JC, Kafatos FC, Goodfliesh R, Hood L (1978a). Proc Natl Acad Sci USA 75:390–394.

Regier JC, Kafatos FC, Kramer KJ, Henrikson RL, Keim PS (1978b). J Biol Chem 253:1305–1314.

Regier JC, Mazur GD, Kafatos FC (1980). Dev Biol 76:286–304.

Rodakis GC (1978). Ph.D. Thesis, University of Athens.

Sim G-K, Kafatos FC, Jones CW, Koehler MC, Efstratiadis A, Maniatis T (1979). Cell 18:1303–1316.

Smith GP (1973). Cold Spring Harbor Symp Quant Biol 38:507–513.

Southern EM (1975). J Mol Biol 98:503–517.

Sures I, Levy S, Kedes L (1980). Proc Natl Acad Sci USA 77:1265–1269.

Tsitilou SG, Regier JC, Kafatos FC (1980). Nucleic Acids Res 8:1987–1997.

Wellauer PK, Dawid IB, Brown DB, Roeder RH (1976). J Mol Biol 105:461–486.

Auxin and Gene Regulation

David C. Baulcombe, Philip A. Kroner, and Joe L. Key
Department of Botany, University of Georgia, Athens, Georgia 30602

I. INTRODUCTION

The mode of action of auxin has been investigated intensely for many years, yet there still is no satisfactory explanation for the molecular basis of any auxin effect. These effects are diverse, including enhancement of cell elongation, stimulation of cell division, modification of cell differentiation, or even delaying cell senescence [Wareing and Phillips, 1978]. It thus seems that auxin may influence everything a plant cell does! It must be a major goal of biology to explain, in molecular terms, how auxin can influence these apparently diverse aspects of plant cell development.

Several models of auxin action have gained due prominence. At one time the observations that auxin effects are associated with, and dependent on, gross changes in RNA metabolism [Key, 1969] led to the notion that auxin-stimulated growth may be mediated by altered gene expression. Subsequently the "acid growth" hypothesis gained favor. This postulates that auxin promotes extrusion of protons into the cell wall, which in turn allows loosening of wall microfibrils such that cell enlargement is facilitated [Cleland and Rayle, 1978]. This loosening may be the result of a direct effect of protons on the microfibrils, or on cell wall hydrolases. Although the "acid growth" hypothesis may be a useful explanation of rapid stimulation of cell elongation by auxin, it throws no light on the mechanisms by which auxin may influence cell division, cell differen-

Levels of Genetic Control in Development, pages 83-97

zone A

zone B

zone C

untreated

20hr auxin
treated

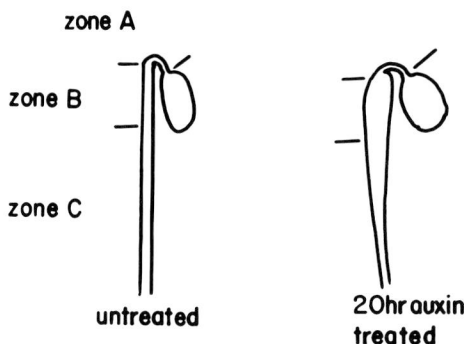

Fig. 1. The effect of 2,4-D on growth of soybean seedlings. Zone A. In untreated tissue this is the meristematic zone where levels of RNA and protein synthesis are high. Following auxin treatment, meristematic activity and macromolecule synthesis is inhibited. Zone B. Normally this is the zone of maximal cell elongation, where the level of polysomes is high. Following auxin treatment, the high level of polysomes is maintained, and RNA synthesis increases severalfold. Cell elongation is suppressed, and radial cell expansion occurs. Zone C. This is the fully differentiated hypocotyl. RNA level is constant and polysomes are low. Auxin activates massive macromolecule synthesis (protein and RNA), and cell division is initiated in the cambial zone.

tiation, or other auxin-regulated processes. It is quite possible that several mechanisms are involved. Vanderhoef has proposed a "dual site" mechanism of auxin action to account for his observations that auxin-stimulated elongation of soybean hypocotyls is biphasic and that the phases display differential sensitivity to auxin analogs, cytokinins, and RNA and protein synthesis inhibitors [Vanderhoef, 1980]. In this report we describe the evidence from a number of sources that auxin action is associated with regulation of gene expression at the pre-translational level. Emphasis is given to the recent work in our laboratory using the intact soybean hypocotyl. This system has a well-defined response to auxin, summarized in Figure 1. This response encompasses stimulation of both cell division and (radial) cell enlargement.

II. AUXIN AND rRNA GENE REGULATION

A well-characterized feature of auxin-stimulated growth in several systems, including the soybean hypocotyl [Key and Ingle, 1967] and artichoke explant culture [Gore and Ingle, 1974], is an increased rate of synthesis of rRNA. In the soybean hypocotyl this is the result of massively increased RNA polymerase I activity, which is thought to result from both an increased RNA polymerase I specific activity and an increased number of RNA polymerase I molecules [Guilfoyle et al, 1975]. In the artichoke explant cultures, in addition to enhanced activity of RNA polymerase I, there is also increased conservation of rRNA sequences during processing of pre-rRNA in auxin-treated artichoke explants [Melanson and Ingle, 1978]. These findings clearly illustrate that auxin-stimulated growth may be associated with an effect at the gene level.

III. AUXIN AND STRUCTURAL GENE REGULATION

Although it is known that the levels of several enzymes increase in auxin-treated tissue, only for cellulase is there any indication that this is due to action at the level of gene expression. Cellulase mRNA levels were observed to change in pea epicotyls (radial cell expansion response) following auxin treatment, when the technique of immunoprecipitation of cellulase peptides from RNA translation products was used [Verma et al, 1975]. Zurfluh and Guilfoyle [1980] have shown that in excised soybean hypocotyls (elongation response) the pattern on 2D gels of newly synthesized proteins is altered following auxin treatment. A similar effect was observed in intact (rootless) hypocotyls [Legocka and Key, unpublished data]. To what extent these pattern changes are the result of altered mRNA levels, rather than modified post-translational processing of peptides, has been investigated with two-dimensional gel analysis of translation products of polyadenylated RNA from auxin-treated and untreated hypocotyls. Representative data are shown in Figure 2, which presents an analysis of translation products from zone B of the hypocotyls. Downward shifts in concentration are observed for five peptides, and upward shifts are observed for five others. Approximately 200 peptides are resolved by this method. In the translation products of total hypocotyl RNA approximately 20 upward shifts and 20 downward shifts are observed [Baulcombe et al, 1980].

A second approach to comparison of polyadenylated RNA of auxin-treated and untreated hypocotyls uses the technique of cDNA/RNA hybridization [Bishop et al, 1974]. Such an analysis revealed that the number and abundance distribution of the polyadenylated RNA sequences is broadly quite similar in both RNA populations (Fig. 2, Table I). The values for the total polyadenylated RNA complexities are in good agreement with the values obtained from soybean tissue culture [Silflow et al, 1979], although they are somewhat higher than the values for polysomal polyadenylated RNA in soybean and other plants [Galau and Dure, in press]. The difference of 7,000 species noted in Table I for the low abundance class is not significant. One minor difference in the hybridization profiles relates to the most abundant sequence class (Table I) in the untreated polyadenylated RNA, which is reduced in concentration in the treated polyadenylated RNA. Comparison of hybridization kinetics with those of ovalbumin mRNA and cDNA (complexity of 1,700 bases) indicates that this highly abundant class of RNA contains only one or two RNA species (Table I) [Baulcombe and

CONTROL

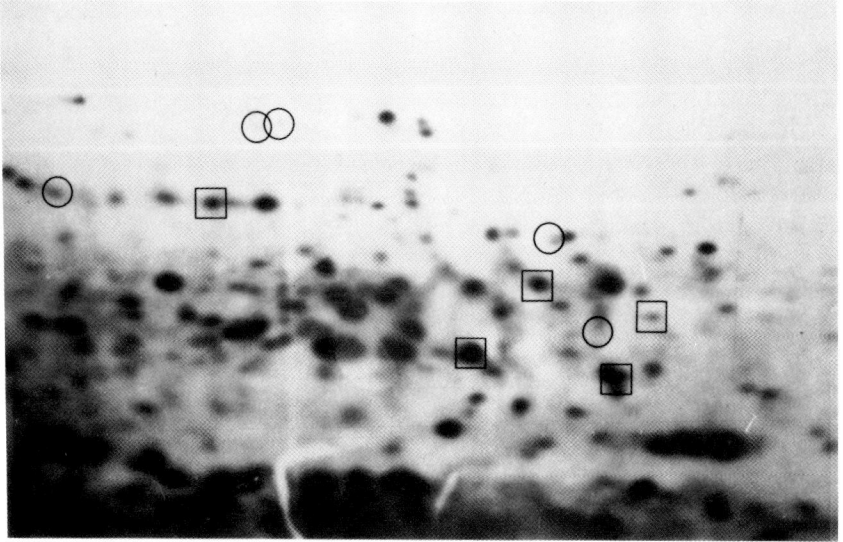

TREATED

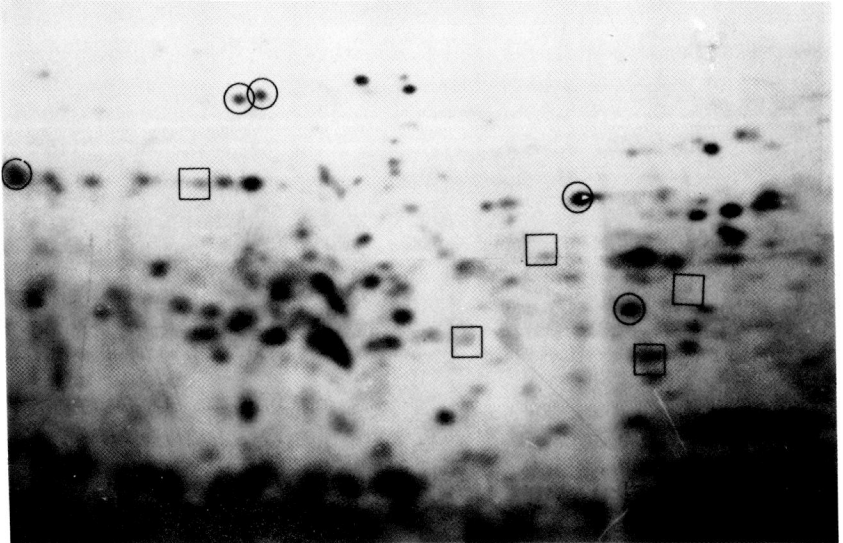

Fig. 2. The effect of 2,4-D on polyadenylated RNA translation products. Total polyadenylated RNA from zone B of hypocotyls was translated in vitro in a wheat germ S-30 system. The ^{3}H-labeled products were analysed by 2D electrophoresis [O'Farrell, 1975] and detected by fluorography [Bonner and Lasky, 1974]. Auxin-increased peptides are circled. Auxin-decreased peptides are surrounded by squares.

TABLE I
Analysis of cDNA Hybridization With Polyadenylated RNA

Transition	P^a	K_{obs}^b	K_{pure}^c	Base sequence complexity (dalton)^d	Number of 1400 nucleotide long sequences^e
Untreated polyadenylated RNA					
1	0.26	0.0035	0.011	1.15×10^{10}	31,681
2	0.19	0.039	0.154	1.0×10^{9}	2,262
3	0.23	0.519	1.692	9.5×10^{7}	205
4	0.09	25.5	218	7.4×10^{5}	2
Auxin-treated polyadenylated RNA					
1	0.27	0.0040	0.009	1.8×10^{10}	38,722
2	0.25	0.060	0.16	1.0×10^{9}	2,178
3	0.14	1.09	5.2	3.1×10^{7}	67

[a]Denotes fraction of cDNA contained in each transition.
[b]The computer-derived rate constant for each transition.
[c]The rate constant for each transition corrected for dilution by other transitions $= \dfrac{\Sigma P}{P} \times K_{obs}$.
[d]Derived by comparison of K_{pure} with K_{pure} ovalbumin. Base sequence complexity (dalton) = (K_{pure} ovalbumin/K_{pure}) $\times$ 1,700 $\times$ 300. K_{pure} ovalbumin under the conditions used was 287.
[e]The number average size of soybean hypocotyl polyadenylated RNA is 1,400 bases [unpublished data].

Key, in press]. Cross-hybridization analyses [Baulcombe et al, 1980] with cDNA probes and mixed hybridizations with a single-copy DNA probe have shown that, apart from this abundant class of sequences, the treated and untreated RNA populations are, within the limits of resolution of the hybridization techniques, largely homologous in terms both of sequence composition and relative sequence abundance. The differences detected in the translation analysis (Fig. 3) [Baulcombe et al, 1980] account for maximally 20 sequences of the 200 or so most abundant sequences in any one cDNA probe (ie, 10% of the abundant class hybridization or 2–5% of the total cDNA). Such differences would be at the limit of resolution of cDNA kinetic hybridization analyses and way beyond the resolution of single-copy DNA hybridizations. Our basic finding, therefore, from in vitro translation analyses and hybridization experiments on polyadenylated RNA, is that approximately 40 different abundant mRNA sequences change in relative concentration following auxin treatment. Sequences are both reduced and increased in relative concentration, but the most prominent effect revealed by cDNA hybridization is the reduction in concentration of about two highly abundant sequences. There is no detectable effect on the low-abundance class of polyadenylated RNA.

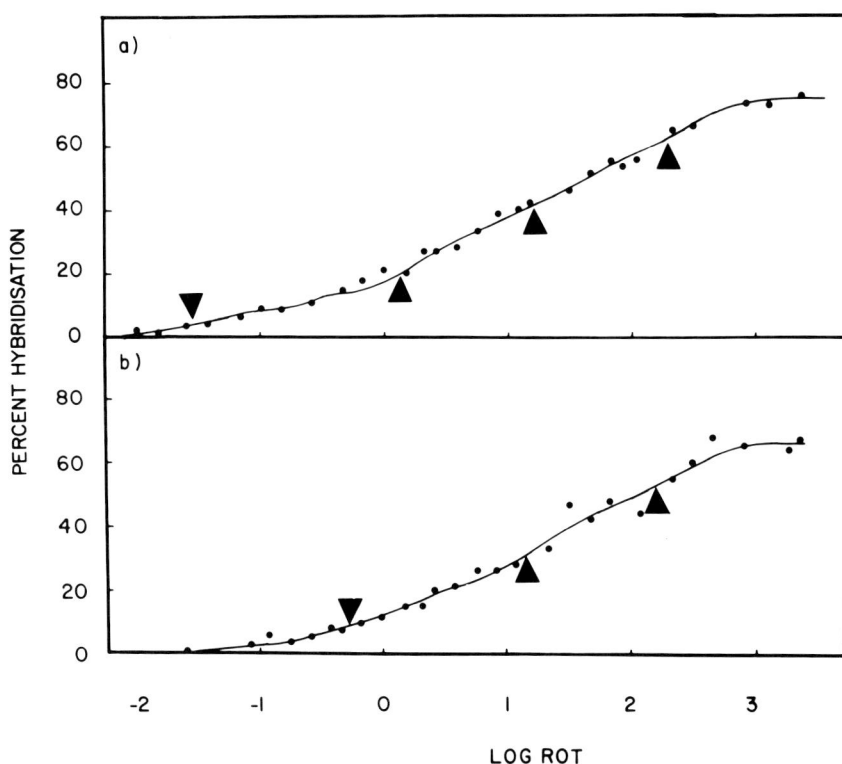

Fig. 3. cDNA/RNA hybridization analysis of polyadenylated RNA from soybean hypocotyls. Total cellular polyadenylated RNA from soybean hypocotyls was hybridized in RNA excess with cDNA. The extent of hybridization was assayed with S1 increase. a) Untreated polyadenylated RNA; b) auxin-treated polyadenylated RNA. Arrows indicate $Rot_{1/2}$ transitions for computer-derived 4(a) or 3(b) component solutions to the data.

IV.   CLONING OF AUXIN-REGULATED POLYADENYLATED RNA

In order to study the molecular basis of gene regulation—in this case auxin-influenced gene regulation—pure sequence hybridization probes are essential. The now widely used recombinant DNA technology has greatly facilitated the isolation of such probes, the main difficulty being the selection of recombinant clones containing the desired sequence. Since the RNA sequences of interest to us had not been prepurified and so could not be used as hybridization probes on a colony hybridization [Grunstein and Hogness, 1975], it was necessary to use a differential colony hybridization procedure. First, recombinant cDNA clones constructed by the GC tail method in pBR322 [Chang et al, 1978] were probed in a colony hybridization with labeled cDNA from the homologous RNA

TABLE II
Properties of Auxin-Regulated Sequences Cloned in pBR 322

Clone number	Inserted cDNA length (bases)[a]	Homologous polyadenylated RNA length (bases)[b]	% of Untreated polyadenylated RNA	% of Auxin-treated polyadenylated RNA[c]	Major peptide[d] product in translation assay (M_r)	Group
1	160	1,000	—	—	33,000	A
2	440	1,000	0.9	<0.02	33,000	A
3	430	1,000	0.7	<0.02	33,000	A
4	410	1,000	0.8	<0.02	33,000	A
5	395	1,000	—	—	33,000	A
6	670	1,000	1.4	0.01	33,000	A
7	375	1,000	1.3	0.02	33,000	A
8	275	350	0.9	0.04	10,000	B
9	330	350	0.8	0.10	10,000	B
10	280	350	1.0	0.20	10,000	B
11	425	750	0.2	0.01	25,000	C
12	380	350	0.9	0.05	10,000	B

[a]Recombinant clones were constructed using the GC tailing procedure such that the inserts could be excised with Pst I [Chang et al, 1978]. Size of inserts was determined by electrophoretic comparison with Alu I digested pBR 322 [Sutcliffe, 1978]. The length of inserted DNA was estimated assuming homopolymer tail length of 30 base pairs. From Baulcombe and Key [in press].

[b]Polyadenylated RNA was electrophoresed in denaturing conditions and the gel was blotted [Alwine et al, 1977] and hybridized to the nick translated clone DNA. The size of the RNA homologous with the clones was estimated by comparison with rRNA, 5S RNA, and tRNA markers. From Baulcombe and Key [in press].

[c]Cloned cDNA sequences were excised with Pst I and nick translated. Strand separation was effected by preparative hybridization [Baulcombe and Key in press], and the single-stranded probe was hybridized in solution with polyadenylated RNA. The concentration of the sequence in the polyadenylated RNA was estimated by comparison with the kinetics of hybridization of ovalbumin RNA and cDNA. From Baulcombe and Key [in press].

[d]Cloned DNA was immobilized on filter discs [Kindle and Firtel, 1978] and hybridized with polyadenylated RNA. The filters were washed and the hybridized RNA eluted off and translated in vitro in a wheat germ S.30. Translation products were analyzed by SDS-PAGE with molecular weight standards.

type. Under the conditions used only clones of abundant sequences produced a signal. These abundant sequence clones were then reprobed using the heterologous cDNA type. Putative auxin-regulated sequences were those that hybridized to either one, but not both, of these probes. Thus 280 of 4,000 cloned untreated RNA sequences hybridized to the homologous (untreated) cDNA. Of these, only 12 did not also hybridize to the treated cDNA [Baulcombe and Key, in press].

Properties of these 12 clones are summarized in Table II. The cDNA inserts are all of different sizes, and therefore they probably are independent. However, the RNA sequences homologous with the cloned sequences fall into three groups: group A comprises RNA(s) of length 1,000 bases, which undergo an approximately 100-fold reduction in concentration following auxin treatment. Seven clones hybridized with this group. Group B contains RNA of 350 bases length, which undergoes a ten-fold reduction in relative concentration following auxin treatment. Four clones hybridized with this group. Group C RNA is 750 bases long and reduced in concentration approximately 15-fold following auxin treatment. There is one clone in this group.

The similarity of the response to auxin by the RNA species within each group suggests that these sequences may be homologous. This is borne out by hybridization/translation experiments [Kindle and Firtel, 1978], which show that the sequences within each group encode electrophoretically similar peptides. Group A encodes a 33 kilodalton peptide, group B encodes a 10 kilodalton peptide, and group C encodes a 25 kilodalton peptide. Further evidence for homology within the groups is shown in Figure 4. Each cloned plasmid was digested with Pst I to excise the insert, electrophoresed in agarose, and blotted. The blotted gels were subsequently hybridized with clones from either of the three groups. The data (Fig. 4) show that cloned cDNA sequences cross-hybridize within their respective groups, but not between groups. We cannot yet exclude the possibility, however, that these groups of RNA species encode closely similar but nonidentical peptides. The repeated isolation of two major groups of auxin-regulated sequences is quite consistent with the cDNA hybridization data (Fig. 2) [Baulcombe and Key, in press], which predict the existence of a small number of auxin-sensitive, highly abundant, polyadenylated RNA sequences in untreated hypocotyls.

V.   THE EXPRESSION OF AUXIN-REGULATED RNA DURING DEVELOPMENT

The strategy of a typical experiment with plant growth substances is to elicit and study the response to treatment with exogenous growth substance. How can one know that the observed effects are in any way related to the action of that growth substance in the intact developing plant? This is difficult to demonstrate

CLONE NUMBER

12 11 10 9 8 7 6 5 4 3 2 1

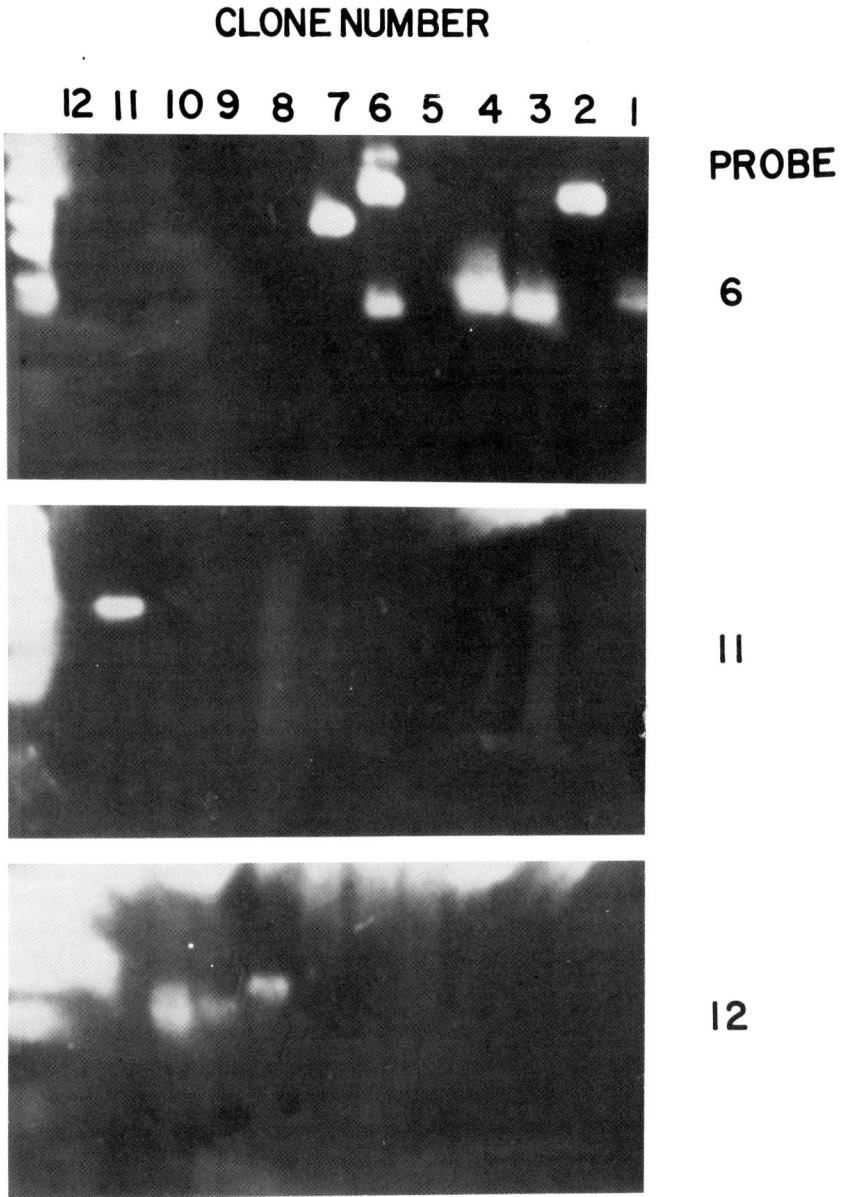

PROBE

6

11

12

Fig. 4. The relationship between group A, group B, and group C cDNA plasmids. Soybean cDNA plasmids 1 through 12 were digested with Pst I, electrophoresed in 2% agarose, and blotted. The blotted gels were subsequently hybridized with nick translated DNA of either group A (clone 6), group B (clone 12), or group C (clone 11) and autoradiographed. The left-most well of each panel shows hybridization of each probe to Alu I digested pBR322. The illustrations are negative images of an autoradiograph.

hrs of auxin

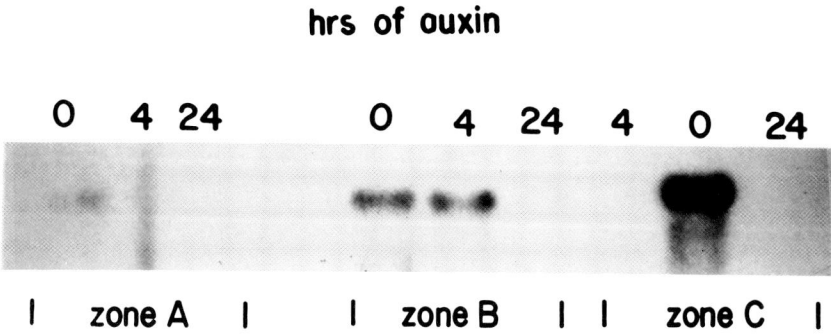

Fig. 5. Developmental expression of group A sequence RNA. Polyadenylated RNA from zones A, B, or C of soybean hypocotyls, which had been treated with auxin for the indicated times, was electrophoresed in formaldehyde/agarose and blotted [Rave et al, 1979]. The blotted gel was subsequently hybridized with nick translated clone 6 DNA (group A) and autoradiographed.

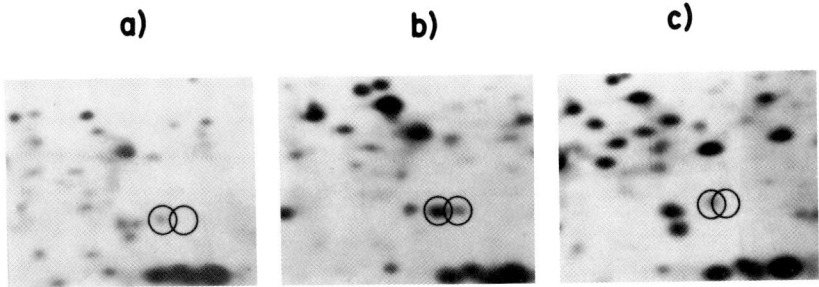

Fig. 6. The relationship between exogenous auxin-regulated and developmentally regulated gene expression. Translation products of polyadenylated RNA from zones A (a) or C (b) of untreated hypocotyls and zone C of four-hour auxin-treated hypocotyls (c) were analyzed on 2D gels and detected by fluorography. These panels illustrate part of a gel in which several peptides appear following exogenous auxin treatment of hypocotyls. In this region of the gel the only auxin-induced change that is mimicked developmentally is the disappearance of two peptides (circled).

directly, so for this work on auxin-regulated gene expression we have had to rely on correlative studies. Auxin levels are highest in the actively growing meristematic regions of the plant and lowest in the differentiated tissue away from these zones. Thus, in the soybean hypocotyl, auxin levels would be highest in zone A and lowest in zone C. For putative auxin-regulated sequences, the relative level should correlate with the auxin level in that zone. The data in Figure 5 show that for the cloned group A RNA sequence, the relative concentration is highest in the zone C and lowest in zone A. In all three zones the level of this sequence is reduced following auxin treatment. A similar gradient in concentration of this sequence, declining toward the auxin-rich meristem, is observed in root tissues (data not shown). The developmental expression of this sequence is therefore quite consistent with auxin being a primary influence in the expression of this sequence. This is not the case, however, for all of the sequences that change in concentration in response to exogenous auxin. Only a fraction of the peptides in the basal hypocotyl region that show a marked response to auxin also show the predicted developmental change in concentration. Part of this analysis is illustrated in Figure 6. Clearly, for these sequences, factors other than auxin can limit the expression of these genes. There is some question then about whether the apparent auxin regulation of such genes has any relevance to normal developmental processes. This doubt should also apply to any exogenous growth substance effect that does not show a predictable developmental shift.

VI. THE MECHANISM OF AUXIN-REGULATED STRUCTURAL GENE EXPRESSION

Evidence concerning the mechanism of auxin regulated gene expression could potentially derive from studies on the auxin receptor or from studies on the gene(s) being regulated. Unfortunately the data on auxin receptors are conflicting. Different laboratories detect nuclear or cytoplasmic receptors or soluble or particulate receptors [Kende and Gardner, 1976]. The nature of auxin receptors must be characterized more completely before we can fruitfully follow this approach.

Our work (following the alternative strategy) using cloned auxin-regulated gene sequences is at an early stage. However, preliminary analysis of nuclear RNA from soybean hypocotyls does suggest that the regulation of those sequences under study is post-transcriptional. Nuclear RNA was prepared from both untreated and auxin-treated hypocotyls and electrophoresed in denaturing gels. The gels were blotted [Alwine et al, 1977] and hybridized against highly labeled DNA. In the total cellular RNA or polyadenylated RNA there is only a single band of hybridization to the cloned auxin-regulated sequence (Fig. 7). In contrast, the nuclear RNA preparations hybridized additionally to higher molecular weight RNA. For clone 6 (group A) the major higher molecular weight component is 1,900 bases long (Fig. 7). Group B nuclear sequences (clone 12) are more heterodisperse. In addition to the 350 base RNA, there is hybridization up to 2,100 bases, with a detectable component at 700 bases [unpublished data]. While there is no precedent from plant systems, such high molecular weight nuclear RNA species are identified in animal cells as mRNA precursors [eg, Ross, 1976].

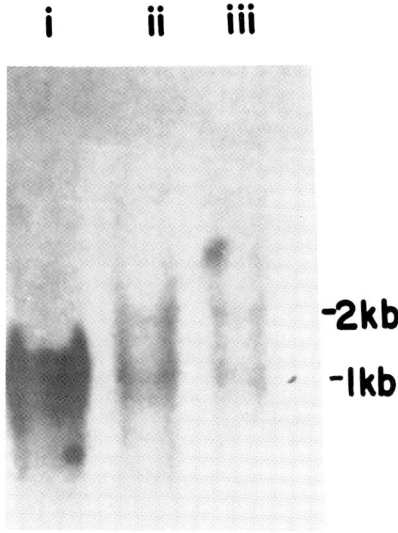

Fig. 7. The effect of auxin on nuclear RNA homologous with auxin-regulated polyadenylated mRNA. Polyadenylated RNA (1 μg) from untreated hypocotyls (i) or nuclear RNA (20 μg) from untreated (ii) or auxin-treated (iii) hypocotyls was electrophoresed on formaldehyde/agarose and blotted [Rave et al, 1979]. The blotted gel was subsequently hybridized with nick translated clone 6 DNA (group A) and autoradiographed.

These putative soybean mRNA precursors are present in nuclei of both auxin-treated and untreated hypocotyls at similar relative concentrations, estimated by intensity of hybridization to the blotted gels (Fig. 6). If the status of these molecules reflects a similar dynamic flow of synthesis and breakdown in both tissues, then the regulation of these sequences must be post-transcriptional. Even the "mature," fully processed 1,000 base and 350 base molecules are detected at similar levels in both treated and untreated nuclear RNAs. This suggests that the regulation may be post-processing, involving perhaps an altered rate of extranuclear transport or decreased stability of these sequences within the cytoplasm. We intend to test these possibilities using cloned copies of RNA sequences that are regulated by auxin in tissue culture, since only in tissue culture can one carry out the kinetic labeling experiments necessary to analyse RNA processing and turnover.

VII.   DISCUSSION AND CONCLUSIONS

It is now well established that eukaryotic cell development is associated with differential gene expression [Galau et al, 1976; Hastie and Bishop, 1976], and plants are no exception [Auger et al, 1979; Kamalay and Goldberg, 1980]. The observations described here and elsewhere [Baulcombe et al, 1980; Baulcombe and Key, in press] indicate that some of the developmental shifts in gene regulation may be influenced by auxin levels. However, the detectable changes in the polyadenylated RNA induced by exogenous auxin applied to soybean hypocotyls are restricted to abundant or moderately abundant sequences [Baulcombe et al, 1980]. In contrast, developmental gene regulation (in tobacco) involves, over the life cycle of the plant, in the order of 50,000 or so low-abundance sequences [Kamalay and Goldberg, 1980]. It would seem, therefore, that either the extensive developmental changes are confined to the non-adenylated mRNA fraction or factors other than auxin regulate these changes.

That the exogenous auxin-influenced changes in gene expression are relevant to normal development processes has been shown by establishment of a correlation between the level of cloned sequences A and B and endogenous auxin level (Fig. 4). Further suggestion that auxin-regulated changes are a significant and integral part of the auxin growth response comes from the timing of these changes. Many of the altered sequence levels, including both increases and decreases, can be detected within four hours of application of auxin [Baulcombe et al, 1980; Baulcombe and Key, in press]. This is relatively early in the context of other auxin-influenced growth changes, including cell division growth, which is not detected until 12 hours after application of auxin [Key et al, 1966]. Probably the best way to assess the significance of any auxin change in gene expression is to correlate the function of the regulated gene with auxin-influenced cellular events. Thus, changes in the expression of genes essential for cell division processes or cell wall biosynthesis, which are both increased in auxin-stimulated tissue, could be considered integral aspects of the auxin growth response. It is likely that many of the peptide changes on 2D gel analyses of translation products fall into this category. However, it is perhaps indicative of the incompleteness of our understanding of cellular aspects of auxin-modified growth that we can make no reasonable guesses as to the identity of genes that are expressed at reduced levels in auxin-treated tissue. This category includes the cloned sequences in groups A–C. In a list compiled by Evans [1974], the only enzyme activity noted to decrease following auxin treatment is auxin oxidase. It is possible that in the near future the identity and function of such unidentified sequences will be investigated by introducing artificial "deformed" auxin-regulated genes into plant cells. The products of such genes might disrupt the function of the intact gene products such that their function can be traced.

The general pattern of eukaryotic gene activity seems to involve constitutive transcription of most sequences [Davidson and Britten, 1979]. Exceptions to this probably include genes of a limited number of terminal differentiation specific peptides, including in vertebrate cells, for example, globin and ovalbumin, and in plant cells ribulose bisphosphate 1,4,carboxylase, and seed-storage proteins.

Although the RNA sequences of cloned groups A–C are expressed at a relatively high level (~1%) in the differentiated basal region of the hypocotyl, they are apparently transcribed constitutively, as evidenced by the presence of a putative pre-mRNA sequence in both auxin-treated and untreated hypocotyls. These genes like the majority of plant genes [Kamalay and Goldberg, 1980], appear to be regulated post-transcriptionally.

We have speculated further, based on data in Figure 7, that the regulation point may be even post-processing, possibly at the level of extranuclear transport or cytoplasmic stability. There are few precedents for this, mainly since most genes being studied are terminal differentiation specific and probably are regulated primarily transcriptionally. However, regulation of globin mRNA stability does occur during DMSO-induced erythropoiesis in murine erythroleukemia cells [Lowenhaupt and Lingrel, 1978]. The next steps in a search for a molecular mechanism of auxin action will involve precise definition of the level of regulation by auxin of gene expression, the identification of the regulatory sequences and the regulator molecules, and a description of the regulation of such regulators. This path should take the course of ever-decreasing circles, with auxin itself at the center.

ACKNOWLEDGMENTS

We thank Cheryle Mothershed for skilled technical assistance. This work is supported by NIH grant CA11624 to J.L.K.

REFERENCES

Alwine J, Kemp D, Stark GR (1977). Proc Natl Acad Sci USA 74:5350–5354.
Auger S, Baulcombe DC, Verma DPS (1979). Biochim Biophys Acta 563:496–507.
Baulcombe DC, Key JL (in press). J Biol Chem.
Baulcombe DC, Giorgini J, Key JL (1980). In "Genome Organization and Expression in Plants." (CJ Leaver, ed.), pp 175–185. Plenum Press, New York.
Bishop JO, Morton JG, Rosbash M, Richardson M (1974). Nature 250:199–204.
Bonner WM, Lasky RA (1974). Eur J Biochem 46:83–88.
Chang ACY, Nunberg JH, Kaufman RJ, Erlich HA, Schimke RT (1978). Nature 275:617–624.
Cleland RE, Rayle DL (1978). Bot Mag Tokyo Special Issue 1:125–139.
Davidson EH, Britten RJ (1979). Science 204:1052–1059.
Evans ML (1974). Annu Rev Plant Physiol 25:195–223.
Galau GA, Dure L (in press).
Galau GA, Klein WH, Davis MM, Wold BJ, Britten RJ, Davidson EH (1976). Cell 7:487–505.
Gore JR, Ingle J (1974). Biochem J 143:107–113.
Guilfoyle TJ, Lin CY, Chen YM, Key JL (1975). Proc Natl Acad Sci USA 72:69–72.
Grunstein M, Hogness D (1975). Proc Natl Acad Sci USA 71:3961–3965.

Hastie ND, Bishop JO (1976). Cell 9:761–774.

Kamalay JC, Goldberg RB (1980). Cell 19:935–946.

Kende H, Gardner G (1976). Annu Rev Plant Physiol 27:267–290.

Key JL (1969). Annu Rev Plant Physiol 20:449–474.

Key JL, Lin CY, Gifford EM, Dengler R (1966). Bot Gaz 127:87–94.

Key JL, Ingle J (1967). In "Biochemistry and Physiology of Plant Growth Substances." (S Wightman and G Setterfield, eds.), pp 711–722. Runge Press, Ottawa.

Kindle K, Firtel RA (1978). Cell 15:763–778.

Lowenhaupt K, Lingrel JB (1978). Cell 14:337–344.

Melanson DL, Ingle J (1978). Plant Physiol 62:761–765.

O'Farrell PH (1975). J Biol Chem 250:4007–4021.

Rave N, Crkvenjakov R, Boedtker H (1979). Nucleic Acids Res 6:3559–3567.

Ross J (1976). J Mol Biol 106:403–420.

Silflow CD, Hammet JR, Key JL (1979). Biochemistry 13:2725–2731.

Sutcliffe JS (1978). Nucleic Acids Res 5:2721–2728.

Vanderhoef LN (1980). In "Genome Organisation and Expression in Plants." (CJ Leaver, ed.), pp 159–173. Plenum Press, New York.

Verma DPS, Maclachlan GA, Byrne H, Ewings D (1975). J Biol Chem 250:1019–1026.

Wareing PF, Phillips IDJ (1978). "The Control of Growth and Differentiation in Plants." Pergamon Press, Oxford.

Zurfluh LL, Guilfoyle TJ (1980). Proc Natl Acad Sci USA 77:357–361.

II [Guilfoyle et al, 1980; Guilfoyle, submitted for publication]. Third, there is a greater increase in template-engaged RNA polymerase I activity than can be accounted for by simply an increase in the amount of the enzyme [Guilfoyle et al, 1980; Guilfoyle, submitted for publication]. It appears that both the rate of RNA chain initiation and chain propagation by RNA polymerase I on the chromatin template increases after auxin application [Guilfoyle and Hanson, 1974; Olszewski and Guilfoyle, 1980].

B. Auxin-Induced Alteration in the Patterns of Translatable Messenger RNA and Protein Synthesis in the Mature Soybean Hypocotyl

Although it is clear from the in vitro studies described and from earlier in vivo studies [reviewed by Key, 1969] that ribosomal RNA synthesis is greatly increased following auxin application to the soybean seedling, these studies give no information about the induction or repression by auxin of specific messenger RNAs or proteins that might be involved in the alteration of growth and development observed in the hypocotyl following hormone treatment. Baulcombe and Key [1980] have shown that the abundance of some polyadenylated RNA sequences is altered following auxin application to the soybean seedling. In addition, Baulcombe et al [1980] have demonstrated by two-dimensional polyacrylamide gel analysis [O'Farrell, 1975] that the products of in vitro translation of polyadenylated RNA are highly different for RNAs isolated from untreated and auxin-treated hypocotyls. As pointed out by Baulcombe et al [1980], in vitro translation followed by analysis of the translation products on two-dimensional polyacrylamide gels permits a relatively detailed estimate of the change in concentration of the most abundant and most efficiently translated messenger RNAs within the total population of RNA.

Although in vitro translation of total or selected fractions of RNA offers a powerful tool with which to study the alteration in the pattern of protein synthesis following auxin application, it is desirable to correlate products translated in vitro with products generated in vivo in both untreated and auxin-treated hypocotyls. A major problem in using intact seedlings for such studies is that in vivo labeling of polypeptides is not adaptable to the short labeling periods required to study rapid auxin effects. First, the process of efficiently applying the radioactive amino acid utilized for labeling the polypeptides is slow since the label would have to be taken up by the roots or rootless seedlings [Guilfoyle, submitted for publication] through the vascular system. Second, the applied radioactive amino acid is greatly diluted by large internal pools of amino acids, which presumably are mobilized from the cotyledons [Guilfoyle, unpublished observation]. To overcome the disadvantages inherent in the in vivo labeling of polypeptides in intact seedlings, we have chosen to work with excised sections of the soybean hypocotyl.

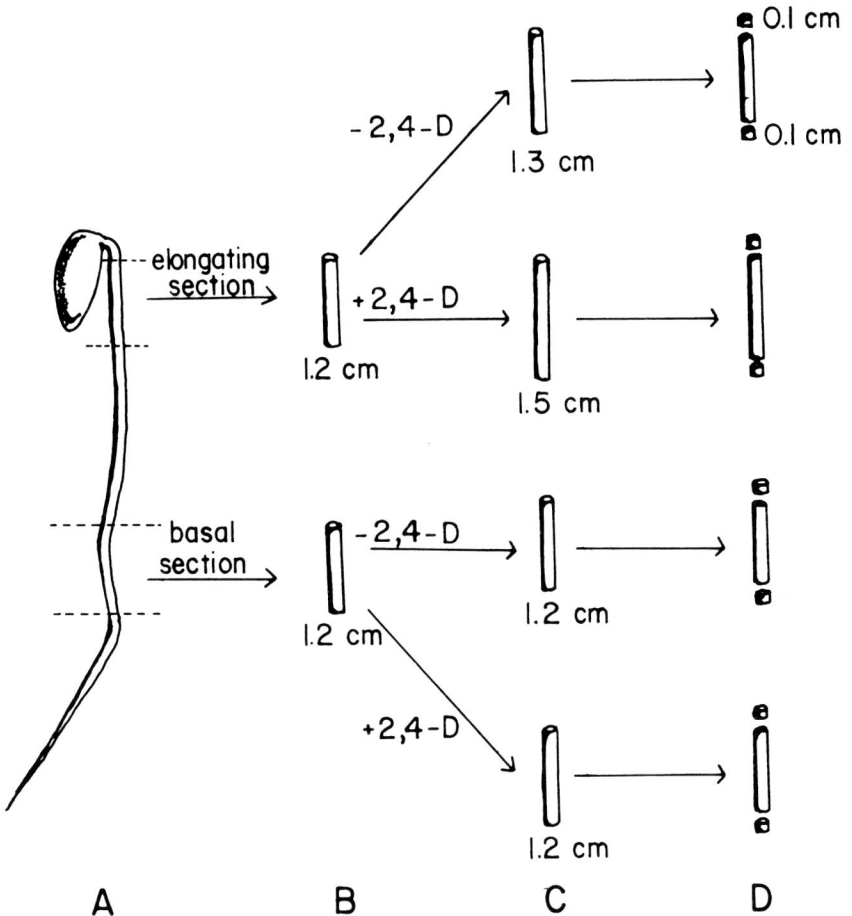

Fig. 2. Diagram of a three-day-old etiolated soybean seedling and of the elongating and basal (mature) sections. A: Three-day-old seedling indicating the zones where sections are excised. B: Elongating and basal sections prior to incubation. C: Elongating and basal sections after a five-hour incubation. The auxin concentration used for incubating elongating sections was 5×10^{-5} M 2,4-D, and that for basal sections was 5×10^{-4} M 2,4-D. Note the difference in section extension in untreated and auxin-treated elongating and basal sections. D: Excision of the terminal 1 mm of incubated sections prior to analysis of radioactive polypeptides. Terminal segments were removed to avoid possible wounding effects on amino acid incorporation into polypeptides.

III. EFFECTS OF AUXIN APPLICATION ON EXCISED BASAL (MATURE) HYPOCOTYL SECTIONS

When basal or mature hypocotyl sections (about 1 cm in length) are excised from three-day-old etiolated soybean seedlings (Fig. 2) and incubated in a buffered solution (pH 6) containing sucrose as an energy source and chloramphenicol as a bacteriostat, there is a rapid increase in the conversion of monoribosomes to polyribosomes from a polyribosome level of 33% to a level of about 50% of the total ribosome population [Travis et al, 1973]. There is also a 70% increase in the protein synthetic activity of 80S ribosomes, which occurs rapidly during incubation [Travis et al, 1973]. These increases are induced by the incubation procedure itself. The amount of RNA in the hypocotyl remains unchanged during incubation [Key and Shannon, 1964], and the template-engaged RNA polymerase I and II activities are relatively unaffected by incubation [Lin et al, 1976]. If 2,4-D (8×10^{-5}M) is added to the incubation mixture, the polyribosome level increases to about 75% of the total ribosome population, and a 120% increase in protein synthetic activity of 80S ribosomes occurs [Travis et al, 1973]. In the presence of 2,4-D, there is a net accumulation of RNA (primarily ribosomal RNA) [Key and Shannon, 1964]. By 12 hours after incubation in auxin, there is a threefold increase in template-engaged RNA polymerase I activity compared to unincubated sections or sections incubated in the absence of 2,4-D, but there is no increase in chromatin-bound RNA polymerase II activity [Lin et al, 1976]. In several respects, then, the excised basal incubated section responds to auxin application in a manner similar to intact basal hypocotyl. Although prolonged incubation of basal sections in the presence of 2,4-D does not result in the pronounced radial cell enlargement and proliferation that is observed in intact observed in intact hypocotyl, the early events induced by auxin appear to be similar in intact and excised sections of soybean hypocotyl. Since the initial responses to auxin application appear quite similar in excised sections and intact hypocotyl, the early events induced by auxin appear to be similar in intact and excised sections of soybean hypocotyl. Since the initial responses to auxin application appear quite similar in excised sections and intact hypocotyl, we believe it is feasible to study the initial events of auxin-induced cell division by using excised sections of basal hypocotyl that are incubated in the presence of 2,4-D. The use of excised sections permits both the rapid uptake of auxin (which is not easy to monitor in an intact seedling) and radioactive label. Both of these are required to study rapid effects of auxin on the in vivo pattern of protein synthesis. In addition, the rapid uptake of auxin is required to study rapid effects on patterns of translated products from in vitro translation experiments.

A. Auxin-Induced Changes in the Pattern of Protein Synthesis Observed by In Vivo Labeling

We have found that proteins can be labeled to sufficient specific activity for analysis on two-dimensional gels by incubating excised basal sections for one to three hours in the presence of radioactive amino acids. Shorter labeling periods are less effective since little of the radioactive label is incorporated into internal regions of the section, and this can be important, especially when analyzing

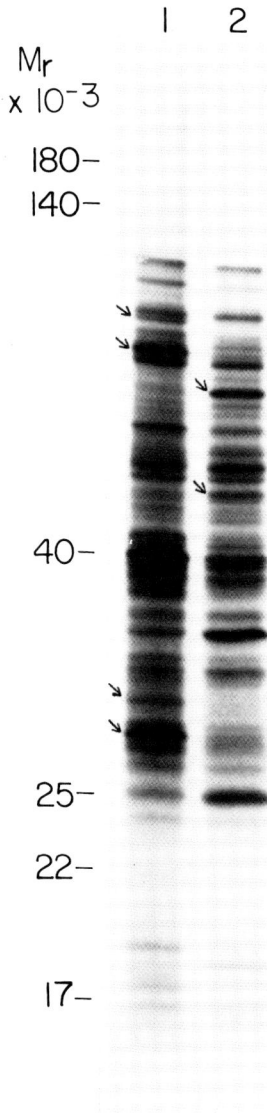

Fig. 3. Fluorograph of in vivo ³⁵S-labeled polypeptides from basal sections resolved by electrophoresis on a concave exponential gradient (10–16%) polyacrylamide gel in the presence of SDS. The labeling period was three hours. Lane 1: Sections incubated in the absence of 2,4-D. Lane 2: Sections incubated in the presence of 2,4-D. Arrows indicate examples of differences in the polypeptides labeled in the absence and presence of auxin.

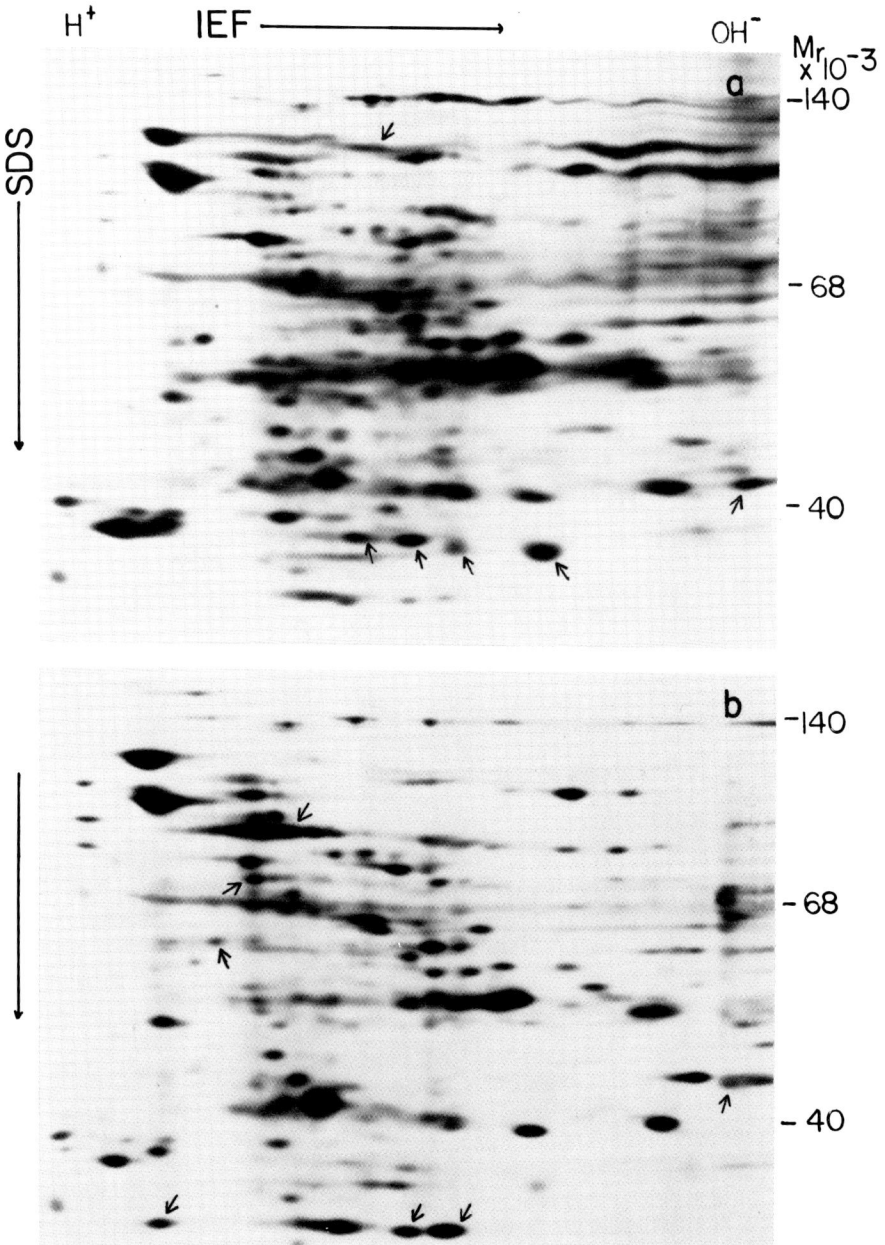

Fig. 4. Fluorographs of in vivo ^{35}S-labeled polypeptides from basal sections resolved by electrophoresis on two-dimensional polyacrylamide gels [O'Farrell, 1975]. Samples were applied to the basic end in IEF. Second-dimension gels are 10% polyacrylamide. The labeling period was three hours. A: Sections incubated in the absence of 2,4-D. B: Sections incubated in the presence of 2,4-D. Arrows indicate examples of auxin-induced and auxin-repressed polypeptides. In all figures of two-dimensional gels presented here, polypeptides on the outer edges of gels are not shown due to variability of polypeptide migration in this region of the gel.

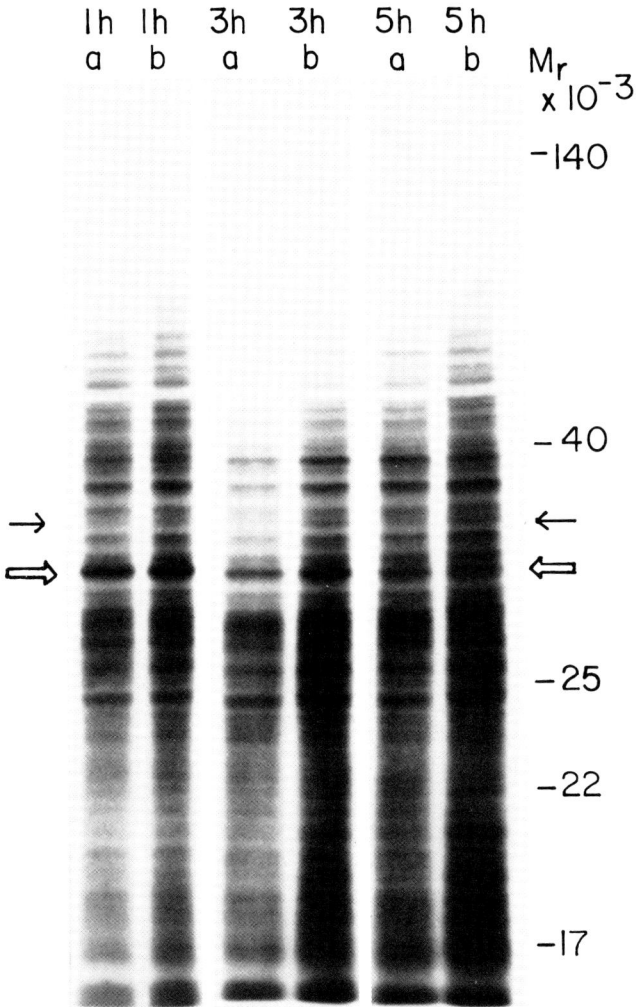

Fig. 5. Fluorograph of in vitro ^{35}S-labeled polypeptides from basal sections resolved by electrophoresis on a 12% polyacrylamide gel in the presence of SDS. Total RNA was isolated from basal sections after incubation by a modification of the method of Lizardi and Engelberg [1979] and was translated in a wheat germ translation system similar to that described by Marcu and Dudock [1974], with ^{35}S-methionine as the radioactive amino acid. Human placental ribonuclease inhibitor isolated as described by Blackburn [1979] was also included in the translation mixture [Scheele and Blackburn, 1979]. A: Sections incubated in the absence of 2,4-D for one, three, and five hours. B: Sections incubated in the presence of 2,4-D for one, three, and five hours. The solid arrow indicates an example of an auxin-induced polypeptide. The open arrow indicates an example of a polypeptide that disappears during incubation of both untreated and auxin-treated sections.

labeled polypeptides in elongating sections [see section IVA; Zurfluh and Guil-foyle, 1980]. Figures 3 and 4 show one- and two-dimensional polyacrylamide gels, respectively, of polypeptides from untreated and auxin-treated excised basal sections of soybean hypocotyl labeled for three hours in the presence of ^{35}S-methionine. Many differences in the pattern of labeled polypeptides are observed on both one- and two-dimensional gels when untreated and auxin-treated polypeptides are compared. It is clear from two-dimensional gel analysis that auxin treatment results in the repression of nearly as many polypeptides as it induces. The labeled polypeptides observed on one- and two-dimensional gels during the three-hour labeling period used here represent the accumulation of labeled polypeptides over this period, and thus the detection of polypeptides depends not only on their synthesis but also on their stabilities. In contrast to this in vivo approach, the in vitro translation of purified messenger RNA would provide a picture of the pattern of polypeptides synthesized at any given point after auxin application. However, this approach makes the assumption that in vitro protein synthesis will display the same pattern of polypeptide synthesis that is actually occurring in vivo. Such an assumption is, of course, not valid in all cases. This approach could nevertheless provide information on how rapidly auxin-induced alterations in polypeptide synthesis can be detected, even if the in vitro translation products cannot be readily correlated with proteins that accumulate in in vivo labeling experiments. In fact, there is evidence that some auxin-induced polypeptides or some polypeptides that are required for continued cell extension may have relatively short lifetimes [Cleland, 1971; Bates and Cleland, 1979; Vanderhoef et al, 1976].

B. Auxin-Induced Changes in the Pattern of Protein Synthesis Observed by In Vitro Labeling

We have purified total RNA from untreated and auxin-treated basal hypocotyl sections that were incubated for one, three, or five hours, and then translated this RNA in a wheat germ translation system using ^{35}S-methionine as the radioactive amino acid. Figures 5 and 6 are one- and two-dimensional gels, respectively, of the translation products from untreated and auxin-treated basal hypocotyl sections. A number of changes in the pattern of protein synthesis can be detected as early as one hour after auxin application (Figs. 5 and 6); however, it is also clear that incubation of the sections, whether in the presence or absence of auxin, also alters the pattern of protein synthesis as analyzed by in vitro translation products.

IV. EFFECTS OF AUXIN APPLICATION ON EXCISED ELONGATING HYPOCOTYL SECTIONS

In intact hypocotyls, cell extension in the elongating zone (Fig. 1.) generally is not enhanced by application of natural or synthetic auxin. In fact, exogenous

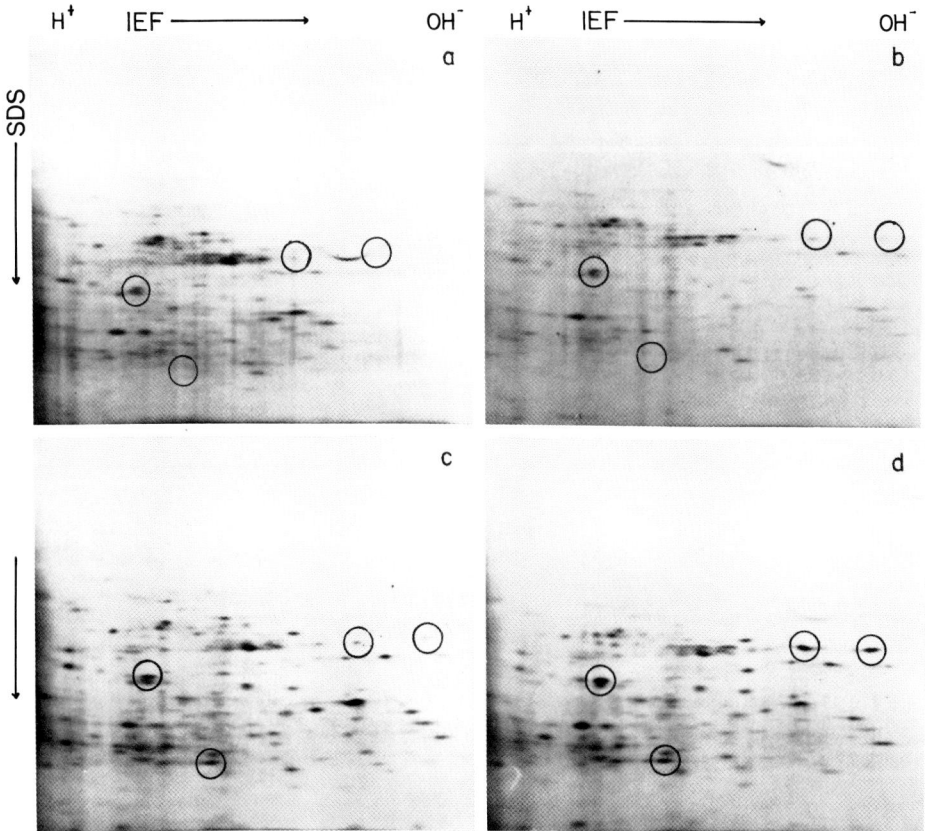

Fig. 6. Fluorographs of in vitro ^{35}S-labeled polypeptides from basal sections resolved by electrophoresis on two-dimensional polyacrylamide gels. Samples were applied to the acidic end in IEF. Second-dimension gels are 10% polyacrylamide. Preparation of RNA and in vitro translation is described in Figure 5. A: Sections incubated in the absence of 2,4-D for one hour. B: Sections incubated in the absence of 2,4-D for three hours. C: Sections incubated in the presence of 2,4-D for one hour. D: Sections incubated in the presence of 2,4-D for three hours. Circles indicate examples of auxin-induced polypeptides.

application of auxin usually results in repressing normal elongation in this zone. Thus it is difficult to study auxin-induced elongation in the intact hypocotyl since the endogenous auxin supply is apparently optimal for cell extension. To circumvent this problem, it has been general practice to excise the elongating portion of the hypocotyl (or epicotyl or coleoptile), whereupon the section becomes dependent on an exogenous source of auxin to elongate normally.

When elongating sections are incubated (Fig. 2), there is a decrease in the level of polyribosomes from 75% to 10–15% of the total ribosome population [Travis et al, 1973], a cessation of net protein synthesis [see discussion by Travis et al, 1973], and a net loss of total RNA [Key and Shannon, 1964]. Under these circumstances, there is a low rate of elongation of this section due to repressed cell extension. When elongating sections are incubated in the presence of auxin, the level of polyribosomes is maintained at nearly the level observed for intact elongation zones [Travis et al, 1973]; net protein synthesis declines but not nearly to the level observed in the absence of auxin [Key, 1964]; and the amount of RNA in the section is maintained at a fairly constant level [Key and Shannon, 1964]. Incubation in the presence of auxin results in a two and one-half to threefold enhancement in the rate of section elongation due to cell extension compared to sections incubated in the absence of auxin.

Auxin-induced cell extension is thought to involve a loosening of inelastic links in the cell wall, which is mediated by the extrusion of H^+ ions across the plasmalemma of the cell into the vicinity of the cell wall [reviewed by Rayle and Cleland, 1977]. Although H^+ extrusion and the initiation of cell extension induced by auxin application can be detected within several minutes after adding the hormone, and although it has been hypothesized that these events are too rapid to be mediated by induced RNA and protein synthesis [Evans and Ray, 1969], recent evidence suggests that both proton extrusion and the initiation of cell elongation require protein synthesis [Rayle and Cleland, 1977; Bates and Cleland, 1979]. It has been known for some time that continued cell extension in the presence of auxin requires both messenger RNA synthesis and protein synthesis [reviewed by Key, 1969], and it now appears that even the rapid induction of cell extension that is observed within minutes after auxin application also requires, at least' protein synthesis.

We are attempting to identify polypeptides that are rapidly synthesized after the addition of auxin to elongating sections of soybean hypocotyl.

A. Auxin-Induced Changes in the Pattern of Protein Synthesis Observed by In Vivo Labeling

We have found that detection of auxin-induced polypeptide synthesis can possibly be masked by wound-induced protein synthesis that occurs at the severed ends of the incubated section [Zurfluh and Guilfoyle, 1980]. We have found it necessary to remove these wounded ends of the section prior to analysis of the polypeptides synthesized in vivo (Fig. 2). Figure 7 shows that wounded ends of the sections synthesize and accumulate a different spectrum of polypeptides than internal portions of the elongating sections. No obvious differences between polypeptides patterns synthesized in the presence or absence of auxin are detected in the ends of the sections when polypeptides are analyzed by one-dimensional

(Fig. 7) or two-dimensional polyacrylamide gel electrophoresis [Zurfluh and Guilfoyle, 1980]. However, when internal portions of the sections are analyzed, differences are observed in the polypeptide patterns synthesized in untreated and auxin-treated sections (Figs. 7 and 8).

In elongating sections of soybean hypocotyl, analysis of auxin-induced changes in the pattern of protein synthesis by in vivo labeling techniques is limited to labeling periods not much shorter than one hour. Elongating sections appear to have a much larger internal pool of amino acids than basal sections, and, thus, the specific activity of the labeled polypeptides is relatively low. In addition, most of the label is incorporated into the ends of sections (about 70% of the label appears in the 1-mm ends of the elongating sections in a three-hour labeling period), and it is the internal portions of the section that must be analyzed if possible wounding effects are to be separated from auxin-induced changes. The greater rate of incorporation of labeled amino acid into the ends of the sections probably results from more rapid penetration of the label into cut ends of the section, where a cuticle does not restrict amino acid uptake; however, it is also possible that wounding of the ends results in a stimulation of protein synthesis in these cells. Since short term in vivo labeling experiments do not appear to be feasible, we are using in vitro translation of total RNA isolated from untreated and auxin-treated elongating sections to analyze early effects of auxin on the synthesis of proteins. Analysis of auxin effects by using in vitro translation is limited only by how rapidly the hormone penetrates the tissue.

B. Auxin-Induced Changes in the Pattern of Protein Synthesis Observed by In Vitro Labeling

Total RNA was purified from internal segments of untreated and auxin-treated elongating sections incubated for one, three, and five hours. The RNA was translated in a wheat germ translation system, and the products were analyzed by one-dimensional (Fig. 9) and two-dimensional (Fig. 10) polyacrylamide gel electrophoresis. Two-dimensional gels (Fig. 10) show a prominent polypeptide of about 30,000 daltons that is synthesized to a much greater amount in the presence of auxin. This polypeptide is detected by one hour after auxin treatment and is also present at three and five hours after incubation in auxin. Several other minor polypeptides appear in only untreated or auxin-treated sections. Since differences are detectable within one hour after auxin application, experiments are now in progress to analyze the spectrum of polypeptides synthesized in vitro at times where auxin-induced cell extension is first detectable (ie, within ten minutes after application of the growth substance).

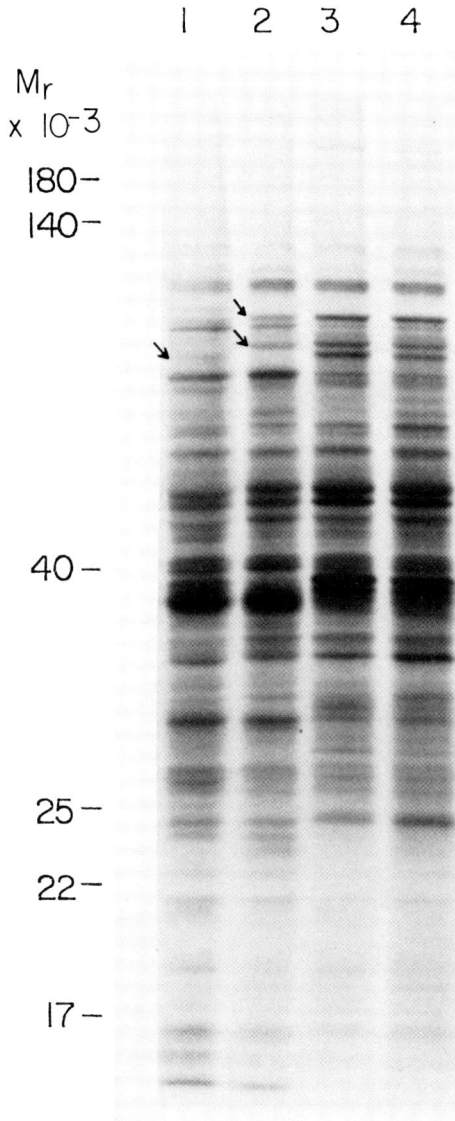

Fig. 7. Fluorograph of in vivo ^{35}S-labeled polypeptides from elongating sections resolved by electrophoresis on a concave exponential gradient (10–16%) polyacrylamide gel in the presence of SDS. The labeling period was three hours. Lane 1: Internal segments of sections incubated in the absence of 2,4-D. Lane 2: Internal segments incubated in the presence of 2,4-D. Lane 3: Terminal 1 mm segments of sections incubated in the absence of 2,4-D. Lane 4: Terminal 1 mm segments of sections incubated in the presence of 2,4-D. Arrows indicate examples of differences in the polypeptides labeled in the absence and presence of auxin.

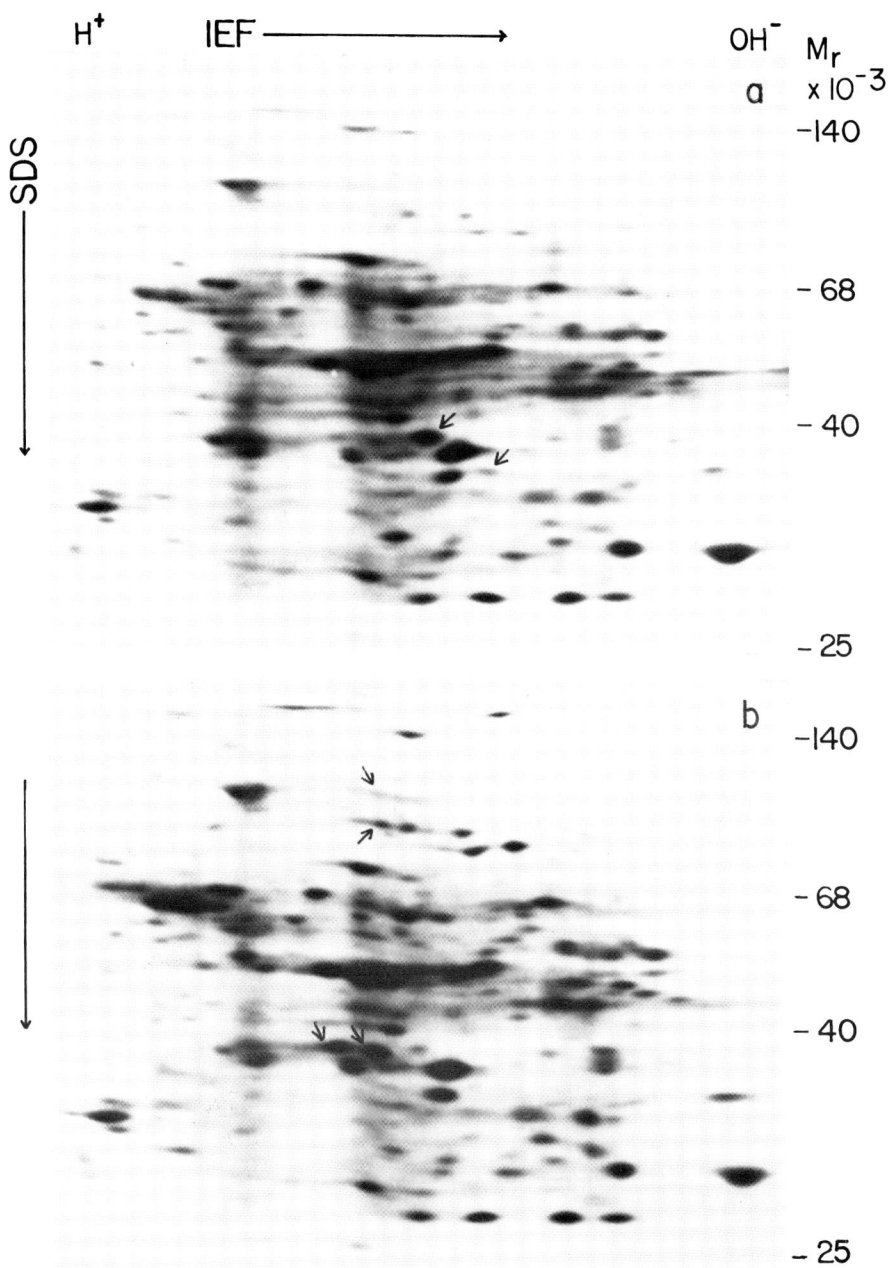

Fig. 8. Fluorographs of in vivo ^{35}S-labeled polypeptides from internal segments of elongating sections resolved by electrophoresis on two-dimensional polyacrylamide gels. Samples were applied to the basic end in IEF. Second-dimension gels are 10–16% exponential gradients. The labeling period was three hours. A: Sections incubated in the absence of 2,4-D. B: Sections incubated in the presence of 2,4-D. Arrows indicate examples of auxin-induced and auxin-repressed polypeptides.

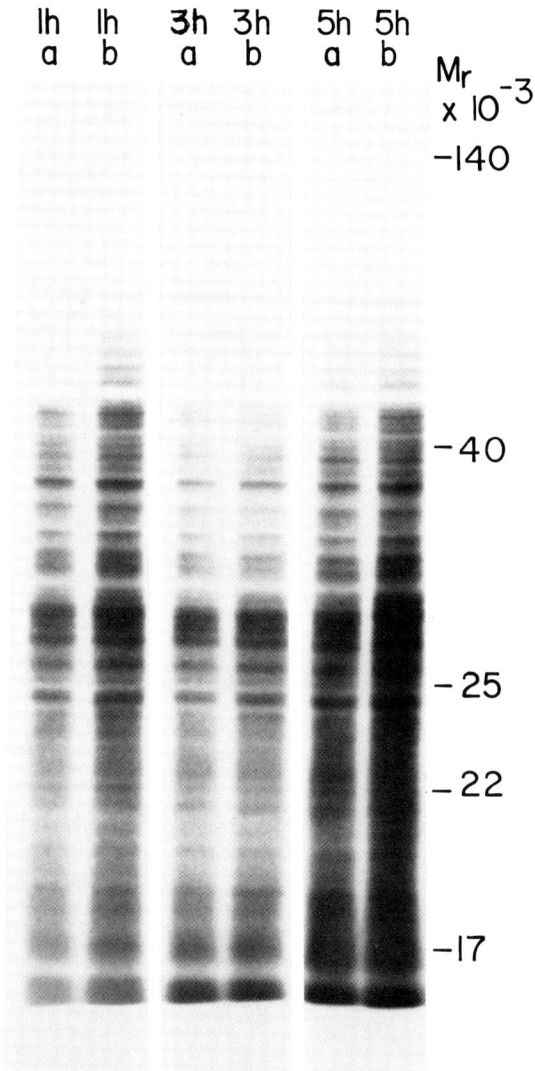

Fig. 9. Fluorograph of in vitro ³⁵S-labeled polypeptides from internal segments of elongating sections resolved by electrophoresis on a 12% polyacrylamide gel in the presence of SDS. Preparation of RNA and in vitro translation is described in Figure 5. A: Sections incubated in the absence of 2,4-D for one, three, and five hours. B: Sections incubated in the presence of 2,4-D for one, three, and five hours.

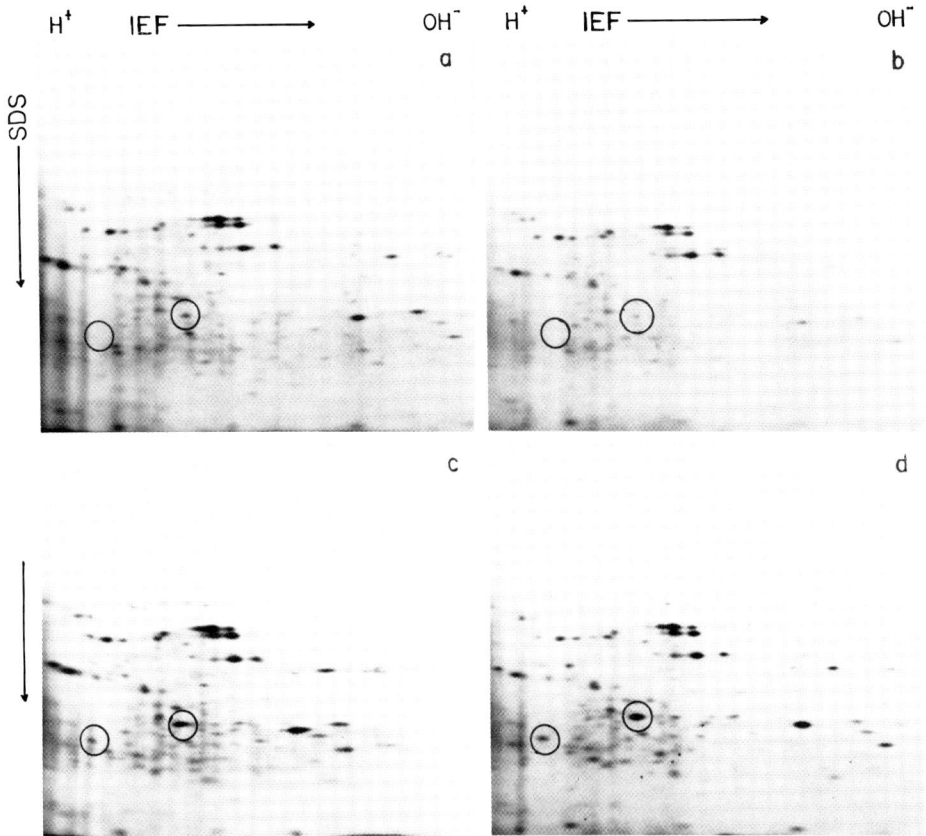

Fig. 10. Fluorographs of in vitro ^{35}S-labeled polypeptides from internal segments of elongating sections resolved by electrophoresis on two-dimensional polyacrylamide gels. Samples were applied to the acidic end in IEF. Second-dimension gels are 10% polyacrylamide. Preparation of RNA and in vitro translation is described in Figure 5. A: Sections incubated in the absence of 2,4-D for one hour. B: Sections incubated in the absence of 2,4-D for three hours. C: Sections incubated in the presence of 2,4-D for one hour. D: Sections incubated in the presence of 2,4-D for three hours. Circles indicate examples of auxin-induced polypeptides.

V.   DISCUSSION AND SUMMARY

We have presented evidence that auxin application to intact and excised mature and elongating sections of soybean hypocotyl alters the pattern of RNA and protein synthesis and accumulation. Although auxin-induced cell division can be studied in intact hypocotyl over periods four to 72 hours after hormone application, rapid effects induced by auxin are not easily studied, and auxin-induced cell extension does not occur in the intact organ. To study rapidly induced effects of auxin application, we have chosen to work with excised hypocotyl sections of different stages of maturity or age. The response to auxin is age specific; the relatively young cells in the elongating zone of the hypocotyl respond to auxin by cell extension, whereas the more mature cells in the basal zone of the hypocotyl respond to auxin by cell division. We believe these responses to be separate from one another, and this is supported by our observations that auxin-induced and auxin-repressed polypeptides are generally different in basal and elongating sections of the hypocotyl. Although it is important to consider both the cell elongation and cell division response to auxin application when developing a hypothesis on the mode of action of this plant growth regulator, it is equally important to keep the two responses in proper perspective. For example, the basal region of the hypocotyl responds to auxin application by producing massive amounts of ribosomal RNA, which is probably a prerequisite for subsequent cellular proliferation; however, the elongating region requires only messenger RNA synthesis and subsequent protein synthesis for auxin-induced cell extension, and the inhibition of ribosomal RNA synthesis does not block this cell extension [reviewed by Key, 1969].

We believe that short-term in vivo labeling experiments and, especially, short-term in vitro translation experiments offer the potential to study rapid alteration of gene expression induced by auxin. Although it can be argued that induced or repressed polypeptide spots on a two-dimensional gel provide little definitive information on the mechanism of auxin action, we can likewise argue that the search for auxin effects on specific enzymes or proteins over the past decades has proved almost fruitless. Even without identifying the function of a particular polypeptide produced by in vitro translation from a population of messenger RNA, techniques are now available to characterize the specific messenger RNA that codes for a particular polypeptide spot [see Baulcombe and Key, 1980]. The use of recombinant DNA technology further offers the potential to identify and isolate the specific genomic DNA sequence that codes for a particular messenger RNA and a particular polypeptide spot. There is then a potential means of determining how auxin application turns on the transcription of specific genes if, in fact, the level of control resides at the transcriptional level.

ACKNOWLEDGMENTS

The authors are indebted to Gretchen Hagen for her critical reading and comments on the manuscript. This research was supported by Public Health Service grant GM24096 from the National Institutes of Health.

REFERENCES

Bates G, Cleland R (1979). Planta 145:437–442.

Baulcombe D, Key JL (1980). J Biol Chem. 255:8907–8913.

Baulcombe D, Giorgini J, Key JL (1980). In "Genome Organization and Expression in Plants." (CJ Leaver, ed.), pp 175–185. Plenum Press, New York.

Blackburn P (1979). J Biol Chem 254:12484–12487.

Chen YM, Lin CY, Chang H, Guilfoyle T, Key J (1975). Plant Physiol 58:78–82.

Cleland RE (1971). Planta 99:1–11.

Evans M, Ray P (1969). J Gen Physiol 53:1–20.

Guilfoyle TJ, Hanson JB (1974). Plant Physiol 53:110–113.

Guilfoyle TJ, Key JL (1977a). In "Nucleic Acids and Protein Synthesis in Plants." (L Bogorad and J Weil, eds.) pp 37–63. Plenum Press, New York.

Guilfoyle TJ, Key JL (1977b). Biochem Biophys Res Commun 74:308–313.

Guilfoyle TJ, Lin CY, Chen YM, Nagao RT, Key JL (1975). Proc Natl Acad Sci USA 72:69–72.

Guilfoyle TJ, Lin CY, Chen YM, Key JL (1976). Biochim Biophys Acta 418:344–357.

Guilfoyle TJ, Olszewski N, Zurfluh L (1980). In "Genome Organization and Expression in Plants." (CJ Leaver, ed.), pp 93–104. Plenum Press, New York.

Gurley WB, Lin CY, Guilfoyle TJ, Nagao RT, Key JL (1976). Biochim Biophys Acta 425:168–174.

Key JL (1964). Plant Physiol 39:365–370.

Key JL (1969). Annu Rev Plant Physiol 20:449–474.

Key JL, Shannon JC (1964). Plant Physiol 39:360–364.

Key JL, Lin CY, Gifford EM, Dengler R (1966). Bot Gaz 127:87–94.

Lin CY, Chen YM, Guilfoyle TJ, Key JL (1976). Plant Physiol 58:614–617.

Lizardi PM, Engelberg A (1979). Anal Biochem 98:116–122.

Marcu K, Dudock B (1974). Nucleic Acids Res 1:1385–1397.

O'Farrell PH (1975). J Biol Chem 250:4007–4021.

Olszewski N, Guilfoyle TJ (1980). Biochem Biophys Res Commun 94:553–559.

Ray PM (1974). In "Recent Advances in Phytochemistry." (VC Runeckles, E Sondheimer, and DC Walton, eds.), vol 7, pp 93–122. Academic Press, New York.

Rayle DL, Cleland R (1977). In "Current Topics in Developmental Biology." (AA Moscona and A Monroy, eds.), vol 11, pp 187–214. Academic Press, New York.

Scheele G, Blackburn P (1979). Proc Natl Acad Sci USA 76:4898–4902.

Thimann K (1977). "Hormone Action in the Whole Life of Plants." University of Massachusetts Press, Amherst, Massachusetts.

Travis RL, Anderson JM, Key JL (1973). Plant Physiol 52:608–612.

Vanderhoef LN (1980). In "Genome Organization and Expression in Plants." (CJ Leaver, ed.), pp 159–173. Plenum Press, New York.

Vanderhoef LN, Stahl C, Lu TY (1976). Plant Physiol 58:402–424.

Zurfluh LL, Guilfoyle TJ (1980). Proc Natl Acad Sci USA 77:357–361.

Organization of the Mitochondrial Genome of Maize

C. S. Levings III and R. R. Sederoff

Department of Genetics, North Carolina State University, Raleigh, North Carolina
27650

I. INTRODUCTION

Mitochondrial genomes constitute an interesting developmental system because they encode information essential for electron transport and oxidative phosphorylation in eukaryotic cells, yet these genomes are small enough that prospects for their detailed analysis may soon be realized. Most of our knowledge of mitochondrial structure and biogenesis comes from studies of fungi and vertebrate cells or tissues. By comparison, little is known about mitochondrial genomes in higher plants. Because mitochondrial genomes of fungi and verte-

Levels of Genetic Control in Development, pages 119–136

brates differ considerably, it is important to know more about the organization and function of mitochondrial genomes in higher plants.

In a wide variety of systems, mitochondrial genomes code for a small number of mitochondrial specific proteins. In mammalian cells or fungi, eight to ten prominent polypeptides are synthesized in mitochondria in the presence of cycloheximide or in isolated mitochondria [Tzagoloff et al, 1979; Mahler and Perlman, 1979; Attardi and Ching, 1979]. Some of these polypeptides are known to be involved directly with respiratory and ATPase protein complexes. These proteins include subunits for cytochrome oxidase, cytochrome b, and a proteolipid component of ATPase. In addition to these polypeptides, mitochondrial genes code for ribosomal RNAs and for a number of specific transfer RNAs [Wood and Luck, 1969; Aloni and Attardi, 1971; Nass and Buck, 1969; Ajai and Borst, 1970].

Of particular interest are mitochondrial coded proteins that are assembled into active complexes with other proteins, which are coded in the nucleus and translated on cytoplasmic ribosomes. In several systems the three largest peptides of the cytochrome oxidase complex are synthesized in mitochondria, and the four lower molecular weight polypeptides are synthesized in the cytoplasm [Weiss et al, 1971; Sebald et al, 1972; Schwab et al, 1972; Mason and Schatz, 1973; Rubin and Tzagoloff, 1973; Koch, 1976]. The coenzyme QH_2-cytochrome reductase complex in yeast and N crassa contains seven or eight polypeptides [Katan et al, 1976a; Weiss and Juchs, 1978], only one of which, the subunit cytochrome b, is coded in the mitochondria. The others are synthesized in the cytoplasm [Weiss, 1972; Weiss and Ziganke, 1974; Katan et al, 1976b; Ross and Schatz, 1976]. The ATPase complex, also known as oligomycin-sensitive ATPase, consists of at least ten protein subunits [Tzagoloff and Meager, 1971]. In yeast four subunits are mitochondrial in origin [Tzagoloff and Meager, 1971]; in Xenopus three products are mitochondrial [Koch, 1976], and in N crassa only two subunits are coded by the mitochondria [Jackl and Sebald, 1975; Sebald, 1977]. One of the ATPase polypeptides, the ATPase proteolipid, is known to be a mitochondrial product in yeast and a cytoplasmic product in Neurospora [Sierra and Tzagoloff, 1973; Sebald, 1979; Sebald et al, 1976].

Among the components used in mitochondria for protein synthesis, a similar interaction of nuclear and mitochondrial components is observed. The ribosomal RNA molecules are coded in the mitochondria, but most of the ribosomal proteins must be coded by the nucleus [Borst, 1972]. Genes coding for a protein of the small ribosomal subunit have been identified in S cerevisiae [Groot et al, 1979] and N crassa [Lambowitz et al, 1976]. In yeast, mitochondrial genes for tRNA have been identified [Martin and Rabinowitz, 1968] that code for all of the common amino acids [Casey et al, 1974; Martin et al, 1977; Martin et al, 1976]. However, aminoacyl synthetases and initiation factors for mitochondrial protein synthesis are themselves synthesized on cytoplasmic ribosomes [Borst, 1972].

II.   MITOCHONDRIAL GENOMES IN HIGHER PLANTS

A. Genome Sizes

All mitochondria perform a common cellular function of oxidative phosphorylation, but mitochondrial genomes vary greatly in size and structure. All animal mitochondrial DNAs (mtDNA) examined have a single duplex circular molecule that contains about 10^7 daltons of DNA (about 15,000 base pairs). In the ciliated protozoa Paramecium and Tetrahymena, the mitochondrial genomes exist as linear molecules of 30×10^6 daltons with unique ends [Cummings et al, 1976; Goldbach and Borst, 1976]. Mitochondrial genomes of fungi are even larger. Yeast mtDNA has been well estimated at 50×10^6 daltons. In Neurospora and Aspergillis, the mtDNA appears to be circular and the molecular weights in these two genera are 40×10^6 daltons and 32×10^6 daltons, respectively [Bernard and Kuntzel, 1976; Lopez Perez and Turner, 1975].

By comparison, the mitochondrial genomes of higher plants are extraordinary in size and range upward from 70×10^6 daltons to 300×10^6 daltons. In one case, the mitochondrial genome of muskmelon has been estimated at 1.1×10^9 daltons [Bendich and Ward, 1980]. Estimates of mitochondrial genome size have been made by electron microscopy (EM), restriction enzyme analysis, and reassociation kinetics (Table I). In spite of the limitations of each kind of estimate, the results, taken collectively, provide strong evidence for exceptionally large and complex mitochondrial genomes in higher plants.

B. Molecular Heterogeneity in mtDNA of Maize

DNA preparations from maize mitochondria contain a sufficient number of circular molecules to permit measurements of molecular size by electron microscopy [Levings et al, 1979; Levings and Pring, 1978]. Instead of a single molecular class, circular molecules of different size are observed. These molecules fall into discrete size classes that vary not only in size but also in relative abundance (Table II). Molecular heterogeneity has been observed in mtDNA for normal (fertile) cytoplasm of maize and in two different types of male sterile cytoplasms. In mitochondria from male sterile cytoplasms S and T, several classes of circular molecules have been observed. Interestingly, each cytoplasmic type had a different distribution of molecular sizes.

C. Molecular Heterogeneity in Other Higher Plants

Evidence indicates that molecular heterogeneity is widely found in higher plants. Early estimates of mitochondrial genome sizes by summation of restriction fragments were larger than the size of the circular molecules observed in

TABLE I
Estimates of Genome Complexity in mtDNA of Higher Plants

Plant	Molecular weight estimate ($\times 10^{-6}$ daltons)	Method of estimation	Reference
Pea	70	Electron microscopy	Kolodner and Tewari [1972]
Pea	74	Renaturation kinetics	Kolodner and Tewari [1972]
Cucumber	120	Restriction digest	Quetier and Vedel [1977]
Potato	90	Restriction digest	Quetier and Vedel [1977]
Potato	99	Renaturation kinetics	Vedel and Quetier [1974]
Wheat	140	Restriction digest	Quetier and Vedel [1977]
Virginia creeper	165	Restriction digest	Quetier and Vedel [1977]
Maize	183	Sum of circles (E.M.)	Spruill et al [1980]
Maize	279	Restriction digest	Levings et al [1979]
Sorghum	100–200	Restriction digest	Pring DR and Conde MF [unpublished]
Muskmelon	1100	Renaturation kinetics	Bendich and Ward [1980]
Flax	106	Restriction digest	Lockhart L and Levings CS [unpublished]

TABLE II
Length and Frequency of Major Classes of Circular
Molecules in Maize mtDNA From Normal Cytoplasm*

Contour length (μm)	Molecular weight ($\times$ 10^{-6} daltons)	Relative abundance (%)
21	45	48
15	33	20
30	66	14
4	8.8	7.6
7	15	5.8
41	91	3.8
9	20	1

*Results shown represent measurements on 104 molecules.
The 9 μm category is represented by only one measured
molecule.

the electron microscope. For example, the sum of the *Eco* RI restriction fragments for the mtDNA from Virginia creeper was 165×10^6 daltons, while the sizes of circular molecules observed in the electron microscope were $60-70 \times 10^6$ daltons [Quetier and Vedel, 1977]. Similar discrepancies were observed in mtDNA from cucumber, wheat, and potato [Quetier and Vedel, 1977]. On the basis of these findings, it was proposed that mitochondrial genomes contained a heterogeneous population of molecules with similar sizes but varied sequence arrangement.

In soybeans, seven discrete classes of circular molecules have been observed [Synenki et al, 1978] with lengths ranging from 6 μm to 29 μm, corresponding to molecular weights of 13 to 64×10^6 daltons, respectively. Similarly, evidence for multiple classes of circular molecules has been obtained in flax [Lockhart and Levings, unpublished results].

D. Possible Explanations of Molecular Heterogeneity

As suggested earlier [Levings et al, 1979], the total information for the normal functioning of plant mitochondria could be coded in more than one molecule. According to this view, each class of circular molecule could represent a discrete chromosome. Distinct multiple chromosomes that differ in sequence have not been directly demonstrated in either mitochondrial or chloroplast genomes. Alternatively, different sizes of molecules could be the result of recombinational events between circular molecules or errors in replication that might alter chromosome size.

III. ANALYSIS OF MITOCHONDRIAL GENOME STRUCTURE IN MAIZE USING CLONED FRAGMENTS OF mtDNA

To investigate the organization of the mitochondrial genome in more detail, *Bam* H1 restriction fragments of normal mtDNA were ligated into pBR 322 and transformed into E coli (LE 392). Specific clones were isolated and identified as mitochondrial in origin by DNA-DNA hybridization using purified mtDNA as labeled probe [Spruill et al, 1980]. Cloned fragments were assigned to specific bands in the total *Bam* H1 digest by co-migration in agarose gels. Cloned fragments were labeled individually with ^{32}P deoxyribonucleotides and hybridized to Southern transfers of *Bam* H1 digests of total mtDNA. In this way, it was possible to determine if a given fragment contained sequences that were shared with other fragments in the mitochondrial genome.

In all cloned fragments examined, hybridization occurred to a band in the digest that co-migrated with the cloned fragment (Fig. I). In these cases the

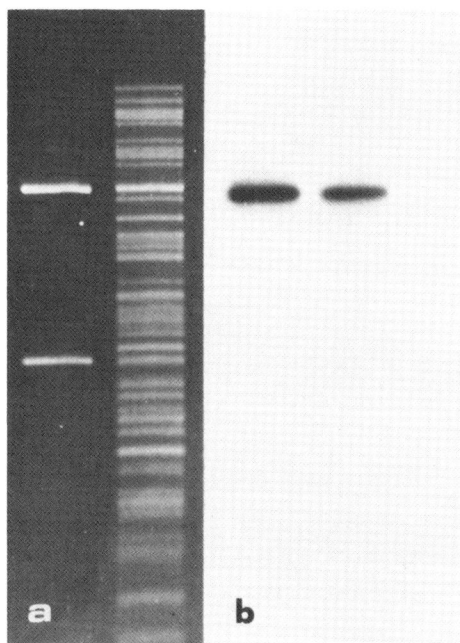

Fig. 1. A cloned *Bam* H1 fragment (629) of mtDNA was digested with *Bam* H1 and run on a 0.8% agarose gel (Fig. 1a, slot 1). Total mtDNA was digested and run in slot 2. Figure 1b shows an autoradiograph following transfer of DNA to nitrocellulose and hybridization against ^{32}P-labeled DNA using purified insert (629). Slots 1 and 2 show hybridizations to DNA from slots 1 and 2 in Figure 1a. The lower band in slot 1, Figure 1a, is plasmid, the upper is inserted mtDNA.

TABLE III
Summary of Results with Cloned Fragments of Normal mtDNA in Maize

Sum of molecular weight of 25 cloned fragments examined	107.9×10^6 daltons
Number of clones that are repeated	4
Molecular weight of repeated sequences (counted once)	11.5×10^6 daltons
Sum of molecular weights of restriction digest of normal mtDNA	183×10^6

[a]Repeated sequences are observed by hybridization of a labeled cloned fragment to a total digest of mtDNA. Repeated sequences are detected when more than the original band that was cloned hybridizes to the probe. [Spruill et al, 1980].

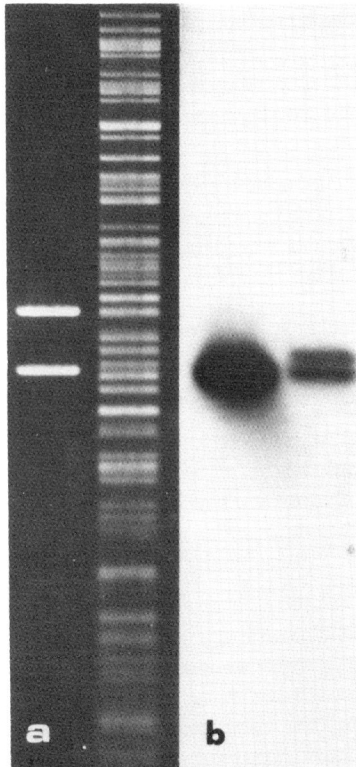

Fig. 2. A cloned *Bam* H1 fragment (178) of mtDNA is shown in an experiment carried out as described in Figure 1. The upper band in slot 1, Figure 2a, is plasmid, and the lower band is the inserted mtDNA. Slots 1 and 2, Figure 2b, show the hybridization of labeled insert to the bands on the Southern transfer of the gel in Figure 2a.

fragment in the digest is considered to be the fragment from which the clone was derived. The total molecular weight of individual clones tested is 108×10^6 daltons, which we estimate to be at least one-third and possibly one-half of the total mitochondrial genome of maize (Table III).

In a small proportion of the fragments (four of 25) hybridization was detected to a second band, indicating some degree of sequence repetition in the mitochondrial genome (Fig. 2). The more significant result, however, is that most fragments do not show homology with other fragments in the digest. If most fragments are unique, it is possible to estimate the size of the mitochondrial genome from the sum of the restriction fragments. Without a measure of shared homology between fragments, restriction digests alone cannot give valid estimates of sequence complexity, since identical sequences may be shared among many fragments.

If each band in the restriction digest is counted once, the total genome size would be 183×10^6. We consider this value to be low because of overlapping bands. In two cases we have identified two clones that hybridize to the same band but are themselves readily distinguished by restriction enzyme analysis. If these fragments thus far tested are representative of the mitochondrial genome, we can estimate the percentage of repeated sequences to be approximately 20%, based on the sizes of the repeated fragments tested and the total sizes of the tested cloned fragments, with each repeat present twice. Consequently, the maize mitochondrial genome is composed largely, but not exclusively, of unique sequences that constitute a very large and complex organelle genome.

IV.   CIRCULAR CLASSES OF mtDNA REPRESENT DISCRETE CHROMOSOMES

The estimate of complexity inferred from the sum of the restriction fragment analysis (183×10^6 daltons) greatly exceeds the size of any individual class of circular molecules found in maize mtDNA. In fact, the sum of the cloned fragments tested directly exceeds the size of each of the classes of circles. Therefore, to possess the number of different sequences observed, the major classes of circular molecules must contain, for the most part, different DNA sequences.

The sum of all the classes of circular DNAs equals 279×10^6 daltons, which is larger than the size of the genome inferred from the sum of the restriction fragments. The discrepancy in this estimate may be due to undetected fragments in restriction digests or to possible monomer–dimer relationships among the major circle classes. Some restriction fragments may be undetected due to overlapping of fragments of similar size, or due to some fragments being present in low abundance. If monomer–dimer relationships existed, the sum of the circular classes would overestimate the complexity. For example, the 45×10^6 dalton and 91×10^6 dalton classes could be related in this way.

It has been suggested that multiple molecular forms of DNA circles (molecular heterogeneity) could be a consequence of recombination or rearrangement of DNA sequences from a single circular class [Levings et al, 1979]. Rearrangement may generate multiple forms of circles as well as increased numbers of restriction fragments. Our results indicate that few fragments share homology and that exchange cannot be the sole basis for molecular heterogeneity. The results strongly support the view that each major class of circular molecule contains different sequences and represents a discrete chromosome.

V. MOLECULAR HETEROGENEITY AS A POSSIBLE MECHANISM FOR CONTROL OF GENE EXPRESSION

If the circular classes contain distinctive sequences and if the circles are present in different abundance, then specific sequences must be present in different relative abundance. Regulation of gene copy number is a well-established mechanism for the control or regulation of gene expression, particularly when gene products are required in large amounts and at specific stages of development. For example, ribosomal RNA gene sequences are present in hundreds of copies, which determine the number of ribosomes synthesized [Brown and Dawid, 1968] and, consequently, the rate of total protein synthesis. If essential DNA sequences are located on different mitochondrial chromosomes, changes in relative abundance of that chromosome could provide an advantage to the organism. This model predicts that those genes whose products are needed and produced in greatest abundance will be found on the most abundant circles.

An appealing aspect of this model is that it may account in part for the large amount of DNA in higher plant mitochondrial genomes. Each specific chromosome could be composed of some essential functional sequences with associated regulatory elements, an origin of replication, and other DNA which is interspersed. Of course, transposable elements, spacers, non-essential regions, and intervening sequences could all contribute to the size of individual chromosomes. In yeast about 50% of the DNA is known to be A–T rich [Prunell and Bernardi, 1974]. A–T-rich sequences as long as 1,000 nucleotides have been reported [Tzagaloff et al, 1979] and have been suggested to serve a spacer or regulatory role. Intervening sequences have been definitely established in mitochondrial ribosomal genes of several fungi and in the genes for cytochrome b and cytochrome c oxidase [Lazarus et al, 1980; Heckman and RajBhandary, 1979; Van Ommen et al, 1980; Alexander et al, 1980; Dujon, 1980].

The presence of multiple chromosomes is typically a feature of eukaryotic nuclear genomes. The circular nature of the mitochondrial molecules, however, is not typical of nuclear genomes. The origin and significance of this unusual feature (multiple chromosomes) of mtDNA of maize is not apparent, but the occurrence in other plants of large mitochondrial genomes and molecular heterogeneity suggests that multiple chromosomes in mtDNA may be a general feature.

VI. DEVELOPMENTAL FUNCTIONS IN THE MAIZE MITOCHONDRIAL GENOME

One obvious explanation for the increased size of mitochondrial genomes in higher plants is that they code for additional functions. Several traits that are unique to higher plants are inherited in an extrachromosomal manner. Reduced kernel size, disease susceptibility, and cytoplasmic male sterility (cms) are among the examples [Levings and Pring, 1979]. Cytoplasmic male sterility, a common trait in higher plants, is inherited in an extrachromosomal fashion and has been reported in at least 80 distinct species [Edwardson, 1970]. In maize, extensive studies have been carried out with cms because of its widespread use in the production of hybrid seed [Duvick, 1965; Laughnan and Gabay, 1975, 1978].

During its development, a normal fertile maize plant produces a tassel which exerts anthers and sheds pollen. The tassels of male-sterile plants do not exert anthers and no pollen is shed. In some strains deformed anthers are exerted which contain aborted pollen grains. Female fertility, as well as other tissues of the plant, are usually unaffected by the cms trait. There is a growing body of evidence that associates the cms trait in maize with the mitochondrial genome [Levings and Pring, 1976, 1979; Pring and Levings, 1978]; however, in other species the chloroplast genome, as well as viral factors, have been suggested.

Genetic analysis has defined several types of cms in maize [Laughnan and Gabay, 1978] by the response of specific cytoplasms to nuclear genes that restore fertility. For example, two dominant genes, Rf1 and Rf2, are required to restore fertility to the Texas cytoplasm (cms-T), whereas a single gene, Rf3, is sufficient to restore fertility to the S cytoplasm (cms-S). The cms-T type is further distinguished by its sensitivity to southern corn leaf blight (Bipolaris maydis race T) and yellow leaf blight (Phyllosticta maydis).

The interaction of the nuclear and cytoplasmic genes that control fertility in maize is strikingly similar to the interaction of nuclear and cytoplasmic mutations in fungi. In both Neurospora [Mitchell and Mitchell, 1956; Bertrand and Collins, 1978] and S cerevisiae [Trembath et al, 1975; Colson et al, 1974] nuclear mutations have been found that suppress specific mitochondrial defects. In Paramecium, Sainsard [1975] has reported a mitochondrial mutation that suppresses a nuclear mutation. Such mutations are expected for genes that involve interacting protein complexes, and they suggest that similar interactions occur in higher plants.

Fig. 3. Gel electrophoresis of mtDNA from *cms-S* maize. Uppermost band contains the large mitochondrial chromosomal DNAs. The lower two bands contain the plasmid-like DNAs, S-1 and S-2, which are found only in *cms-S* types.

VII. THE S CYTOPLASM OF MAIZE

A. S Cytoplasm Contains Unusual Plasmid-Like DNAs

Among the distinctive types of cytoplasmic male sterility that exist in maize [Duvick, 1965], perhaps the most interesting is the S cytoplasm, which can be distinguished from other types on the bases of 1) specific nuclear genes, called restorers of fertility (Rf), which suppress the expression of male sterility [Duvick, 1965; Laughnan and Gabay, 1978], 2) electron microscopy and restriction enzyme fragment analysis of mtDNAs [Pring and Levings, 1978; Levings et al, 1979b], 3) the nature of mitochondrial protein products [Forde and Leaver, 1980], and 4) the presence of two plasmid-like mtDNAs (S-1 and S-2) (Fig. 3), which are uniquely associated with *cms-S* [Pring et al, 1977]. These plasmid-like DNAs designated S-1 and S-2, have molecular weights of 4.1 and 3.4 ×

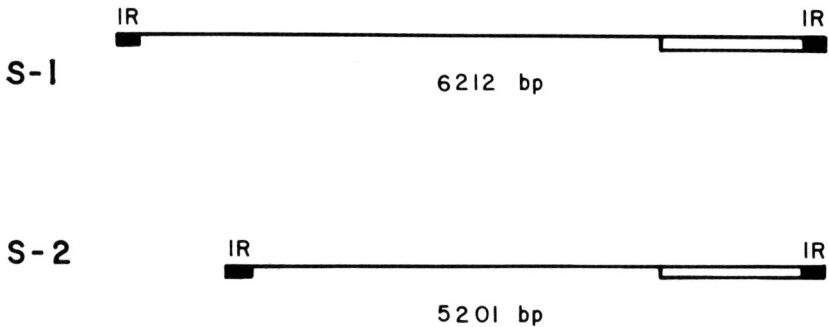

Fig. 4. Diagrammatic representation of the plasmid-like DNAs, S-1 and S-2. Sequence homology between S-1 and S-2 is marked by rectangles below lines. Position of terminal inverted repeats is designated by solid areas in rectangles. S-1 and S-2 contain 6212 and 5201 base pairs (bp), respectively.

10^6 daltons, respectively. Electron microscopy has indicated that both the S-1 and S-2 molecules exist in a linear configuration; this is in marked contrast to the mitochondrial chromosomal DNAs, which are observed as covalently closed circles [Levings et al, 1979]. These unique DNAs have been successfully isolated only from mitochondrial preparations; attempts to isolate these molecules from chloroplast and nuclear preparations have failed. Finally, strict maternal transmission has been demonstrated for the S-1 and S-2 DNAs.

The S-1 and S-2 DNA molecules associated with S cytoplasm have not been found in mtDNA preparations from maize with normal (fertile) cytoplasms or in any of the other male-sterile cytoplasms, T, C, EP, ES, BB. This finding has been repeatedly confirmed with many different accessions of these cytoplasms. Furthermore, the plasmid-like DNAs have been encountered in every source of the S cytoplasm examined regardless of nuclear background. These initial observations suggest a relationship between the plasmid-like DNAs and the S type of male sterility.

B. Sequence Organization of S-1 and S-2

The S-1 and S-2 DNA molecules have interesting sequence arrangements (Fig. 4). Both DNA molecules are terminated by inverted repeats that are approximately 200 nucleotides in length [Levings and Pring, 1979] and that constitute about 3% of the length of the molecules. This feature was demonstrated by configurations formed when these DNAs were denatured and self-annealed. Furthermore, heteroduplexing studies show that the S-1 and S-2 DNAs contain 196 nucleotides in common at one end of the molecules and 1,347 nucleotides in common at the other end. These findings indicated that S-1 and S-2 each contain about 1,550 base pairs of sequence homology, which are terminally located and include the inverted repeats. Although the importance of these terminal inverted repeats is not clear, it is instructive to remember that in lower organisms inverted repeats are often prominently involved with insertional events [Heffron et al, 1975; Sheldrick and Berthelot, 1975; Bukhari et al, 1977].

Although most of the various cytoplasmic male-sterile types are very stable, this has not been the case with cms-S. Laughnan and his colleagues have encountered spontaneous mutations in which cms-S individuals have reverted to fertiles [Laughnan and Gabay, 1978]. Through genetic analysis, they established that two distinctive kinds of changes were occurring, those in which the change takes place in the cytoplasm and those in which it happens in the nucleus. Revertant strains of the first type arise most frequently and behave in subsequent generations as though there has been a permanent change at the cytoplasmic level. Based upon an analysis of the spontaneous revertants, they proposed the existence of a male fertility element that has the characteristics of an episome [for greater detail see Laughnan and Gabay, 1978].

C. Reversion of cms-S to Fertility

Comparison of the mtDNAs of cms-S and its cytoplasmic revertants has revealed important distinctions [Levings et al, 1980]. In this study the cytoplasmic revertant strains arose from cms-Vg (M825/Oh07), which is one of a number of independently discovered male-sterile types falling in the general category of cms-S. The strain cms-Vg (M825/Oh07) is especially inclined to the cytoplasmic reversion event [Laughnan and Gabay, 1978]. When mtDNAs of a standard cms-S and cms-Vg (M825/Oh07) are compared, both contain the S-1 and S-2 plasmid-like DNAs. In cms-S, S-1 and S-2 are observed in equimolar quantities, but in cms-Vg (M825/Oh07) S-2 is present in reduced amounts. It may be that the decreased quantity of S-2 DNA detected in mitochondrial preparations of this source is correlated with the cytoplasmic instability so frequently encountered in this strain.

The more astonishing discovery is that the S-1 and S-2 DNAs are no longer detectable in mtDNA preparations from cytoplasmically reverted strains. Thus far, this has proved to be the case with every revertant strain examined. Although traces of S-1 and S-2 may be present in the revertants, there is no question that the disappearance of the plasmid-like DNAs is coincident with the phenotypic change from cytoplasmic male sterility to male fertility.

Restriction endonuclease fragment analysis with Xho 1 of mtDNA from cms-S, cms-Vg (M825/Oh07) and its revertants has uncovered further distinctions. Differences were not observed between cms-S and cms-Vg (M825/Oh07) with regard to number and position of bands on electrophoretograms. However, most

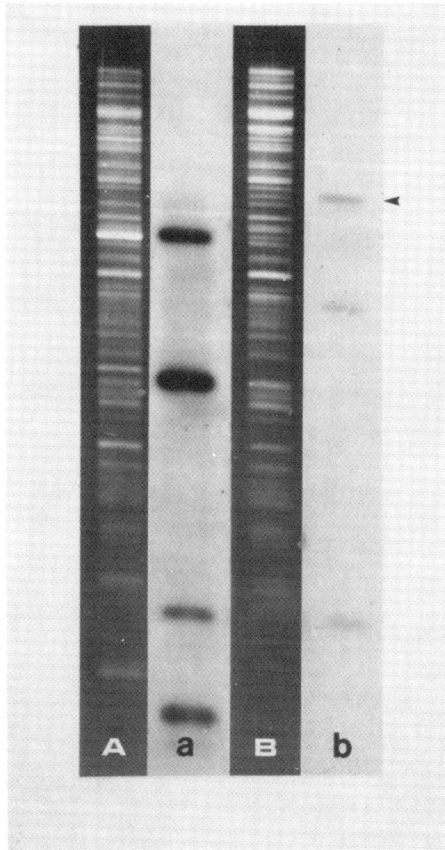

Fig. 5. Gel electorphoresis of *Xho* I digests of mtDNAs (A, B) and autoradiographs of corresponding Southern blots hybridized with ^{32}P-labeled S-2 DNA, respectively (a, b). MtDNAs are from (A,a) *cms-Vg* and (B,b) a fertile revertant. Arrow indicates a hybridization coinciding with a newly arisen band in the revertant. This figure is intended to illustrate one example of mtDNA alterations associated with reversion from male sterility to fertility.

revertant strains exhibited new fragment bands that were not seen in the *cms-S* and *cms-Vg* (M825/Oh07) control strains (Fig. 5). Loss of the plasmid-like DNAs coincident with the appearance of altered mtDNAs suggests the possibility that revertant strains may have originated by the insertion of S-1 and/or S-2 DNAs into the mitochondrial chromosomal DNAs. This possibility was tested by DNA–DNA hybridization studies with Southern transfers and labeled S-1 and S-2 DNA probes. Hybridization with labeled S-2 DNA probes identified homology among unique bands in fertile revertant strains, which were not detected in the sterile controls (Fig. 5). Importantly, several hybridizations coincided with the new restriction fragments observed in the *Xho* I restriction patterns. This outcome showed that some of the new bands do indeed carry sequences that are homologous with the S-2 plasmid-like DNA. In addition, the hybridization studies indicated several regions of homology common to the sterile control and the fertile revertant strains. This result demonstrates that sequences homologous to S-2 DNA exist in the mitochondrial chromosomal DNA of both the sterile and the revertant strains.

Similar hybridization studies with labeled S-1 DNA and Southern blots of *Xho* digests also disclosed several regions of homology common to the sterile and fertile revertant strains, as well as a few differences in homology among them [Levings et al, 1980]. Several hybridization bands seemed in common with both the S-1 and S-2 probes; however, this was not surprising in as much as S-1 and S-2 contain sequence homology of approximately 1,500 base pairs (Fig. 3). Interestingly, the hybridization bands produced with S-1 probe did not unambiguously match the new *Xho* I fragments of the fertile revertants. Apparently, those sequences of the S-2 plasmid-like DNA that are not common with S-1 are more prominently concerned with the origin of the new bands.

These results indicate that the reversion of the *cms-Vg* individuals to the fertile condition is correlated with the disappearance of free plasmid DNAs, S-1 and S-2, as well as alterations in the mitochondrial chromosomal DNAs of the revertants. Furthermore, hybridization studies have implicated the plasmid-like DNAs with these changes. It is apparent that rearrangements have taken place in the mitochondrial chromosomes that prominently involve the plasmid-like DNAs. Even though mutations or other events cannot yet be dismissed, it is attractive to speculate that the insertion of plasmid-like DNA sequences into the mitochondrial chromosomes is associated with the reversion to male fertility. This explanation is strengthened by the diversity observed in the mtDNAs among the various revertants, since these distinctions would be in accord with insertional events occurring at different sites in the mitochondrial chromosomes. Finally, the S-1 and S-2 molecules may be uniquely prepared for transposition because both DNAs have terminal inverted repeats.

D. Relationship of S-1 and S-2 to Transposable Elements

Transposable elements have been recognized for years in the nuclear genome of maize, where they exert abnormal control over the activities of standard genes [for a review see Fincham and Sastry, 1974]. In her pioneering studies, Mc-Clintock [1956, 1961] chose the term "controlling elements" to describe these elements, and to date three distinct classes have been identified on the basis of their specificity: Dotted (Dt), Activator (Ac), and Supressor-mutator (Spm). The present data suggest that there may exist in maize an additional class of transposable elements that consists of the two elements S-1 and S-2 and prevails in the mitochondria. Unlike the "controlling elements" of the nuclear genome, which may effect many different loci, the S-1 and S-2 elements seem to influence only the male fertility trait. This does not preclude the possibility that these elements could be inserted near or into other mitochondrial genes and thereby modify their expression. However, our primitive understanding of the mitochondrial genome suggests that such events would be undetected, if indeed they survived. Certainly, the S-1 and S-2 DNAs, which are 6,212 and 5,201 base pairs in size, respectively, are large enough to code for a number of genes. In this sense, the S elements may be analogous to transposable elements found in prokaryotic systems that carry such other genes as antibiotic resistance, which is not responsible for the insertional or excisional events [for a review see Bukhari et al, 1977]. Finally, although the relative importance of the S-1 and S-2 elements remains unknown, the apparent involvement of two components is strikingly similar to that of the *Ac* and *Spm* systems of the nuclear genome.

Even though the reversion of the S cytoplasm from the male-sterile to the male-fertile condition correlates well with changes in the mitochondrial genome, there is an indication that a nuclear influence is involved. In certain nuclear backgrounds, cytoplasmic reversions to the fertile condition are observed far more frequently than in others [Laughnan and Gabay-Laughnan, 1979]. For example, in the inbred *M825* nuclear background, fertile revertants, either partial or whole tassel, occurred as frequently as 10% of the time, whereas in WF9, reversions were not detected. Furthermore, results of the present study suggest that nuclear background may quantitatively influence the mitochondrial DNA, S-2. Clearly, interactions between the nuclear and mitochondrial genomes are indicated.

ACKNOWLEDGMENTS

We wish to thank W. M. Spruill, Jr., for permission to reproduce autoradiographs for Figures 1 and 2. This work was supported by grants from USDA/SEA (5901-0410-9-0356-0), NSF (PCM 76-09956-A01), and NIH (7-R01-GM-26267-02).

REFERENCES

Aaji C, Borst P (1970). Biochim Biophys Acta 217:560–562.
Alexander NJ, Perlman PS, Hanson DK, Mahler HR (1980). Cell 20:199–206.
Aloni Y, Attardi G (1971). J Mol Biol 55:271–276.
Attardi G, Ching E (1979). In "Methods in Enzymology." (S Fleischer and L Packer, eds.), vol 56, pp 66–79. Academic Press, New York.
Bendich A, Ward BL (1980). In "Genome Organization and Expression in Plants." (CJ Leaver, ed.), pp 17–30. Plenum Press, New York.

Bernard U, Kuntzel H (1976). In "The Genetic Function of Mitochondrial DNA." (C Saccone and AM Kroon, eds.), pp 105–109. North Holland, Amsterdam.

Bertrand H, Collins RA (1978). Mol Gen Genet 166:1–13.

Borst P (1972). Annu Rev Biochem 41:333–376.

Brown DD, Dawid IB (1968). Science 160:272–280.

Bukhari AI, Shapiro JA, Adhya SL (eds) (1977). "DNA Insertion Elements, Plasmids and Episomes." Cold Spring Harbor Laboratory, Cold Spring Harbor, New York.

Casey JW, Hsu HJ, Getz GS, Rabinowitz M (1974). J Mol Biol 88:735–747.

Colson AM, Goffeau A, Briquet M, Weigel P, Mattoon JR (1974). Mol Gen Genet 135:309–326.

Cummings DJ, Goddard JM, Maki RA (1976). In "The Genetic Function of Mitochondrial DNA." (C Saccone and AM Kroon, eds.), pp 119–130. North Holland, Amsterdam.

Dujon B (1980). Cell 20:185–197.

Duvick DN (1965). Adv Genet 13:1–56.

Edwarson JR (1970). Bot Rev 36:341–420.

Fincham JRS, Sastry GRK (1974). Annu Rev Genet 8:15–50.

Forde BG, Leaver CJ (1980). Proc Natl Acad Sci USA 77:418–422.

Goldbach RW, Borst P (1976). In "The Genetic Function of Mitochondrial DNA." (C Saccone and AM Kroon, eds.), pp 137–142. North Holland, Amsterdam.

Groot GSP, Grivell LA, van Harten-Loosbrook N, Krieke J, Moorman AFM, van Ommen GJB (1979). In "Structure and Function of Energy Transducing Membranes." (K van Dam and BF van Gelder, eds.), North Holland, Amsterdam.

Heckman JE, RajBhandary UL (1979). Cell 17:583–595.

Heffron F, Rubens C, Falkow S (1975). Proc Natl Acad Sci USA 72:3623–3627.

Jackl G, Sebald W (1975). Eur J Biochem 54:97–106.

Katan MB, Pool L, Groot GSP (1976a). Eur J Biochem 65:95–105.

Katan MB, van Harten-Loosbrook N, Groot GSP (1976b). Eur J Biochem 70:409–417.

Koch G (1976). J Biol Chem 251:6097–6107.

Kolodner R, Tewari KK (1972). Proc Natl Acad Sci USA 69:1830–1834.

Lambowitz AM, Chua NH, Luck PJL (1976). J Mol Biol 107:223–253.

Laughnan JR, Gabay SJ (1975). In "Genetics and Biogenesis of Mitochondria and Chloroplasts." (CW Birky, Jr, PS Perlman, TJ Byers, eds.), pp 330–349. Ohio State University Press, Columbus.

Laughnan JR, Gabay SJ (1978). In "Maize Breeding and Genetics." (DB Walden, eds.), pp 427–446. John Wiley and Sons, Inc., New York.

Laughnan JR, Gabay-Laughnan, S (1979). Maize Genet Coop Newsletter 53:90–92.

Lazarus CM, Lunsdorf H, Hahn U, Stepien PP, Kuntzel H (1980). Mol Gen Genet 177:389–397.

Levings CS III, Pring DR (1976). Science 193:158–160.

Levings CS III, Pring DR (1978). In "Stadler Symposium." vol 10, pp 77–94. University of Missouri Press, Columbia.

Levings CS III, Pring DR (1979). In "Physiological Genetics." (JG Scandalios, ed.), pp 171–193. Academic Press, New York.

Levings CS III, Shah DM, Hu WWL, Pring D, Timothy DH (1979). In "Extrachromosomal DNA," ICN-UCLA Symp Mol and Cell Biol, (DJ Cummings, P Borst, IB Dawid, SM Weisman, and CF Fox, eds.), vol XV, pp 63–73. Academic Press, New York.

Levings CS III, Kim BD, Pring DR, Conde MF, Mans RJ, Laughnan JR, Gabay-Laughnan SJ (1980) Science. 209:1021–1023.

Lopez Perez MJ, Turner G (1975). FEBS Lett 58:159–163.

Mahler HR, Perlman PS (1979). In "Extrachromosomal DNA," ICN-UCLA Symp Mol and Cell Biol (DJ Cummings, P Borst, IB Dawid, SM Weisman, and CF Fox, eds.), vol XV, pp 11–33. Academic Press, New York.

Martin NC, Rabinowitz M (1968). Biochemistry 17:1628–1634.

Martin NC, Rabinowitz M, Fukahara H (1977). Biochemistry 21:4672–4677.

Martin R, Schneller JM, Stahl AJC, Dirheimer G (1976). In "Genetics and Biogenesis of Chloroplasts and Mitochondria." (T Bucher, W Neupert, W Sebald, and S Werner, eds.), pp 755–758. North Holland, Amsterdam.

Mason TL, Schatz G (1973). J Biol Chem 248:1355–1360.

McClintock B (1956). Cold Spring Harbor Symp Quant Biol 21:197–216.

McClintock B (1961). Am Natur 95:265–277.

Mitchell MB, Mitchell HK (1956). J Gen Microbiol 14:84–89.

Nass MMK, Buck CA (1969). Proc Natl Acad Sci USA 62:506–513.

Pring DR, Levings CS III (1978). Genetics 89:121–136.

Pring DR, Levings CS III, Hu WWL, Timothy DH (1977). Proc Natl Acad Sci USA 74:2904–2908.

Prunell A, Bernardi G (1974). J Mol Biol 86:825–841.

Quetier F, Vedel F (1977). Nature 268:365–368.

Ross E, Schatz G (1976). J Biol Chem 251:1997–2004.

Rubin MS, Tzagoloff A (1973). J Biol Chem 248:4275–4279.

Sainsard A (1975). Nature 257:312–314.

Schwab AJ, Sebald W, Weiss H (1972). Eur J Biochem 30:511–516.

Sebald W, Weiss H, Jackl G (1972). Eur J Biochem 30:413–417.

Sebald W, Graf T, Wild G (1976). In "Genetics and Biogenesis of Chloroplasts and Mitochondria." (T Bucher, W Neupert, W Sebald, and S Werner, eds.), pp 167–174. North Holland, Amsterdam.

Sebald W (1977). Biochim Biophys Acta 463:1–27.

Sebald W (1979). Eur J Biochem 93:587–599.

Sheldrick A, Berthelot N (1975). Cold Spring Harbor Symp Quant Biol 39:667–678.

Sierra MF, Tzagoloff A (1973). Proc Natl Acad Sci USA 70:3155–3159.

Spruill WH Jr, Levings CS III, Sederoff RR (1980). Dev Genet. 1:363–378.

Synenki RM, Levings CS III, Shah D (1978). Plant Physiol 61:460–464.

Trembath MK, Monk BC, Kellerman GM, Linnane AW (1975). Mol Gen Genet 140:333–337.

Tzagoloff A, Meager P (1971). J Biol Chem 246:7328–7336.

Tzagoloff A, Macino G, Sebald W (1979). Annu Rev Biochem 48:419–441.

Van Ommen GJB, Boer PH, Groot GSP, De Haan M, Roosendaal E, Grivell LA, Haid A and Schweyen RJ (1980). Cell 20:173–183.

Vedel F, Quetier F (1974). Biochim Biophys Acta 340:374–387.

Weiss H, Sebald W, Bucher Th (1971). Eur J Biochem 22:19–26.

Weiss H (1972). Eur J Biochem 30:469–478.

Weiss H, Ziganke B (1974). Eur J Biochem 41:63–71.

Weiss H, Juchs B (1978). Eur J Biochem 88:17–28.

Wood DD, Luck DJL (1969). J Mol Biol 41:211–224.

Maize Mutants Altered in Embryo Development

William F. Sheridan and M. G. Neuffer

Department of Biology, University of North Dakota, Grand Forks, North Dakota 58202, and Department of Agronomy, University of Missouri, Columbia, Missouri 65211

I. INTRODUCTION

Embryonic development in plants must involve the regulated expression of a genetic program. Just as in mammals, insects, nematodes, slime molds, and other species, the genesis of the complex organism involves a precise sequence of gene activities during the development of a single cell (a zygote or spore) into the multicellular individual. So, it seems reasonable that similar genetic mechanisms must underlie the formation of the precise patterns of cells that are

Levels of Genetic Control in Development, pages 137–156

produced by division of the zygote, and its cellular progeny, during plant embryogenesis.

Among animals, particularly Drosophila [Gehring, 1976; Baker, 1978] and mouse [McLaren, 1976], numerous mutants are known that alter the normal sequence of embryogenesis. In many cases, these mutants are known characteristically to suffer from a developmental block at a specific stage of embryogenesis. The abundance of such mutants and the analysis of their chromosomal locations and allelic relationships indicate the complexity of the genetic contribution to embryonic development. Furthermore, the identification and genetic analysis of these mutants sets the stage for their biochemical characterization. In the long run, therefore, the isolation and study of mutants affecting embryogenesis should contribute to a more profound understanding of the molecular basis of embryonic development.

II. MAIZE EMBRYO AND ENDOSPERM DEVELOPMENT

The most extensive analysis of maize embryo development was carried out by Randolph [1936] on the dent variety "Pride of Michigan" at Ithaca, New York. Subsequently, Wang [1947] reported on the embryological development of inbred and hybrid maize embryos at Urbana, Illinois; Kiesselbach [1949] reported on embryo development in a midwestern dent variety at Lincoln, Nebraska; Abbe and Stein [1954] studied the development of the shoot apex during embryogeny of the inbred A188 at St. Paul, Minnesota; and Paxson [1963] investigated the effects of some plant growth regulators on embryogeny of the hybrid of the Texas lines 401 and 403 at College Station, Texas.

From these studies, there has emerged a detailed description of maize embryo development. There are two variables that can exert a considerable influence on the rate of embryo development: 1) the environmental conditions under which the maize plants are grown, with a more rapid rate occurring in warmer, more southern locales than in northern ones, and 2) the genotype of the strain being examined. The following description of maize embryo development is for field-grown maize in Columbia, Missouri. It is based largely on an analysis of the above-cited reports plus our own observations using F_2 materials from crosses of genetic stocks with a large embryo strain (Alexander's High Oil).

Randolph [1936] and Kiesselbach [1949] identified the stage of embryo development on the basis of the number of days since pollination. Abbe and Stein [1954] used a system based on morphological stage and utilized the appearance of leaf primordia for characterization of the later stages of development.

Fertilization of the egg nucleus occurs about 15 to 24 hours after pollination, depending on the temperature and silk length. Cell division occurs during days 2 to 5 after pollination to produce an elongated, club-shaped *proembryo*. The apical region of the proembryo contains smaller, more dense, meristematic cells and subsequently develops into the embryo proper. The basal portion of the proembryo comprises the suspensor, which ceases growth early and has an unknown function in subsequent embryonic development. The proembryo reaches a length of 0.2 to 0.4 mm.

Following proembryo development, there is a two-day period, termed the *transition phase* by Abbe and Stein [1954] (see Fig. 1), during which the apical region of the proembryo begins to differentiate. An epidermis forms, and a region on the upper anterior face of the proembryo begins to divide rapidly. Internally, the longitudinal axis of the embryo begins to form as a result of more rapid cell divisions. These divisions lead eventually to the formation of the shoot-root meristem. This axis, which arises at an oblique angle to the longitudinal axis of the proembryo, determines the subsequent orientation of the developing embryo. The transition phase ends with the appearance of bulges on the anterior surface of the embryo.

These bulges develop into the coleoptile and the shoot apex. At the same time, a marked swelling on the posterior side appears, which, along with the uppermost region of the embryo, develops into the scutellum. This is termed the *coleoptilar stage* by Abbe and Stein. During this period, which lasts about two days and ends at about nine days postpollination, the embryo, including the scutellum, has continued to grow and reaches a length of about 1 mm.

The bulges above and below the shoot apex fuse to form a ring of tissue that develops into the coleoptile, which covers, except for a small opening (the coleoptilar pore), the developing shoot apex throughout the subsequent stages of development. These stages are marked during the early period, from about day 11 to about day 21 or somewhat later, by the rapid growth of the scutellum. This structure quickly enlarges to form a large, shield-shaped structure on the posterior side of the embryo, which extends above the stem apex and eventually wraps part way around the anterior surface of the embryo to partially enclose the coleoptile. During this period and subsequently, the shoot apical meristem gives rise to the leaf primordia, which sequentially develop until five or six primordia have usually formed by the time the seed is mature (45 to 60 days depending on growth conditions). These stages are numbered 1 through 6 by Abbe and Stein (Fig. 1) according to the time of appearance of the six leaf primordia. Note that the number of days accompanying each stage in Figure 1 corresponds to the development rate observed in Minnesota, where summer field conditions may result in a somewhat slower growth rate than that observed in Missouri. At maturity the embryo, including scutellum, is about 8 mm long and constitutes about 9–13% of the weight of the mature kernel.

Endosperm formation begins a few hours after fertilization, with the embryo sac rapidly becoming filled with endosperm nuclei while the embryo has reached only the 10 to 24 cell stage. After three or four days, the endosperm forms cell

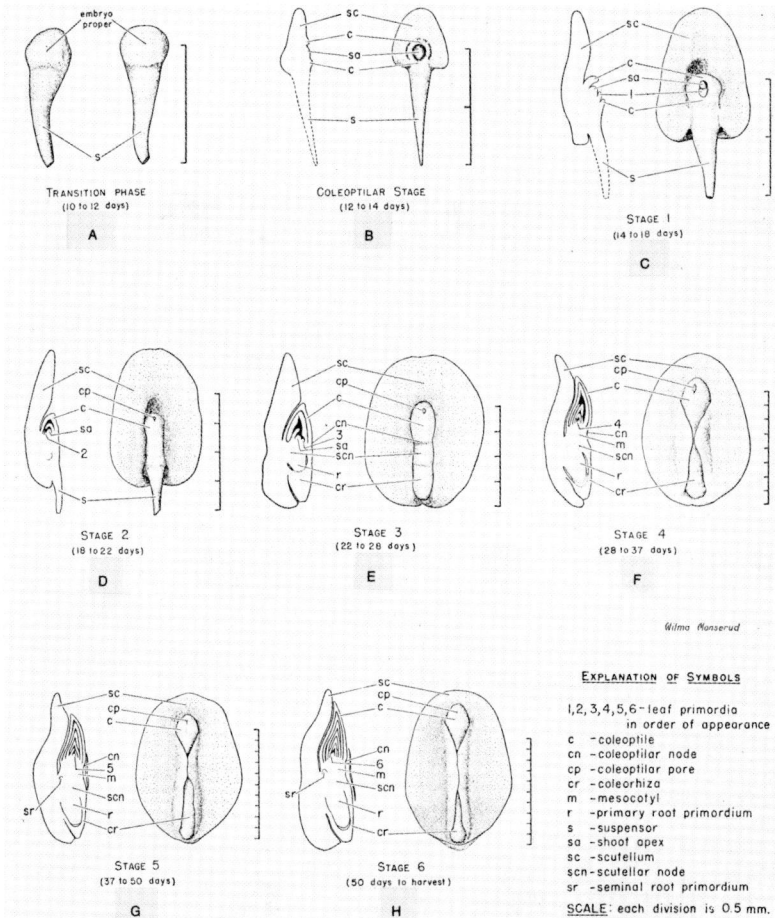

Fig. 1. Maize embryo stages according to Abbe and Stein [1954]. A. Transition phase; side view to the left, face view to the right. B–H. Coleoptilar stage to stage 6, respectively: in each set the face view of the embryo (right) is accompanied by a slightly diagrammatic median sagittal section (left) of the same embryo. *Note:* the coleoptilar pore is omitted from the median sections because it commonly fails to fall in precisely the median plane as established with reference to the shoot apex. Reproduced from Abbe and Stein [1954] with permission.

walls and passes from the free nuclear stage to the cellular stage. During its development, the endosperm changes tremendously in size. Randolph [1936] reported that the length of the endosperm was 0.3 mm at two days after pollination, was 2 mm at eight days, 10 mm at 24 days, and 15 mm at 48 days. At the same time that it is undergoing this great enlargement, the endosperm changes its shape from a narrow-based pyramid with the embryo at the base to a much larger structure, narrower at the base, with much of its bulk in the upper half of the seed. Wang [1947] reported that at maturity, the endosperm of individual kernels of two inbred lines and their reciprocal hybrids ranged in dry weight from 199 to 240 mg, while the embryos from these kernels had a dry weight of from 17 to 23 mg each. The large embryo strain that we have used varied somewhat from these dimensions.

III. DEFECTIVE KERNEL MUTANTS

Among flowering plants, a rather limited number of mutants, scattered among several different species, are known to affect the embryonic sequence. The defective kernel mutants of maize, wherein both the endosperm and embryo are abnormal, have been known for more than 50 years. In his pioneering work, Mangelsdorf [1926] described 14 non-allelic mutants of this type and reported on their retarded embryonic development. Although a few similar reports appeared at about the same time [Brink, 1927; Wentz, 1925; Emerson, 1932], and two more reports were subsequently published [Lowe and Nelson, 1946; Brink and Cooper, 1947], little has been done with them since the original report. Yet, these types of mutants should be of considerable value in the study of the genetic control of embryonic development. A recent pair of papers pointed out the value of lethal mutants for studying plant embryogenesis [Meinke and Sussex, 1979a] and reported the characteristics of four such mutants in Arabidopsis [Meinke and Sussex, 1979b].

The defective kernel mutants, because of the defectiveness of both the endosperm and the embryo, may seem less suitable for the genetic analysis of embryogenesis than those mutants wherein only the embryo is defective (see the germless mutants of Demerec [1923] and Wentz [1930]). Although the germless mutants are intriguing and undoubtedly deserve study, there are certain advantages in the use of the defective kernel mutants for studying embryo development. There are many more defective kernel mutants than germless mutants. The defective kernel mutants allow for the study of embryo–endosperm interaction in a way that is not possible with the germless mutants since the endosperm of the latter is nearly normal in phenotype [Neuffer and Sheridan, 1980]. Furthermore, if we restricted our investigation to those mutations affecting only the embryo, we would be eliminating from consideration the largest class of lethal embryo mutants—those which have a gene-regulated defect for a process or a substance vital for both endosperm and embryo development. The doubly defective mutants allow us to identify and characterize mutants of a nutritional nature.

Fig. 2. Mature ear (upper) and immature ear 16 days post-pollination (lower) showing the 3-to-1 segregation pattern expected after self-pollination. The defective kernels are readily recognized on both ears because of their smaller size, and on the immature ear they are lighter colored and more translucent than the normal kernels.

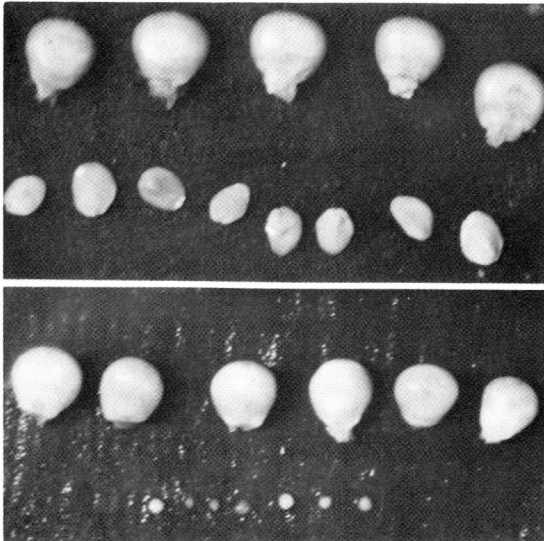

Fig. 3. Normal kernels and embryos (upper) and mutant kernels and embryos (lower) from the same self-pollinated ear 16 days post-pollination. Note that although the mutant kernels are reduced in size compared to the normal kernels, the embryos are much smaller than their normal counterparts.

The greatest advantage of the use of defective kernel mutants has become apparent during our study of them. Because these mutants become distinguishable at an early point in kernel development (Fig. 2), the mutant kernels and embryos (Fig. 3) can be readily studied over a large portion of their developmental sequence. This allows for culturing of immature embryos, as well as their dissection, morphological description, and histological examination [Sheridan and Neuffer, 1980].

All of these studies are readily carried out with a large number of mutants because the defective nature of their endosperm makes these mutants easily distinguishable from normal kernels from an early stage onward. Consequently, we have been able to examine many more mutants and more gene loci in greater detail than we would have if we had limited our study to *germless*-type mutants.

IV. METHODS OF ISOLATION AND ANALYSIS OF MAIZE MUTANTS WITH ALTERED EMBRYOGENESIS

All of the mutants included in this study were induced by treatment of pollen with ethylmethane sulfonate in paraffin oil. The pollen was crossed onto normal silks and the resulting kernels were planted and grown as an M_1 generation. The mutants were selected by screening ears of self-pollinated M_1 plants. Three criteria were applied: Ears were selected that showed a good 3:1 segregation of normal to mutant kernels; the kernel phenotype was clearly defective and severe; and the mutant kernels were inviable (lethal) when tested for germination in a sand bench [Neuffer and Sheridan, 1980]. The locating of mutants to chromosome arm was carried out by the method of Roman and Ullstrup [1951] using the B-A chromosome set of J. B. Beckett. A detailed description of the technique appears in the report by Beckett [1978], and additional details on our use of it have been presented by Neuffer and Sheridan [1980].

Immature, normal and mutant kernels were collected from segregating ears. A portion of each sample was dissected in order to determine embryonic stage and to weigh the embryo and endosperm. The embryos were classified for embryonic stage using the system of Abbe and Stein [1954]. Usually, this was done on the basis of morphology of dissected embryos; but in some cases, sectioned material was used. In many cases, both normal and mutant kernels were fixed, embedded in paraffin, sectioned, and stained using the tertiary butyl alcohol method and safranin-hematoxylin essentially as described by Sass [1968].

Normal and mutant embryos from immature kernels were cultured on a basal medium (4% sucrose and the mineral salts of Murashige and Skoog [1962]), as well as on an enriched medium consisting of the basal medium supplemented with 20 amino acids, seven vitamins, and five nucleic acid bases. Additional details on media composition, culture techniques, and methods of evaluation of culture results are contained in Sheridan and Neuffer [1980].

V. DISTINGUISHING TWO CLASSES OF DEFECTIVE KERNEL MUTANTS

In our studies of a collection of 150 recessive defective kernel mutants [Neuffer and Sheridan, 1980; Sheridan and Neuffer, 1980], we have come to distinguish two major groupings on the basis of operational definitions: nutritional-type mutants and developmental-type mutants. The majority of the mutants have immature embryos that grow to some degree, often to a rather large size and advanced stage of development, but die before kernel maturity. These mutants are lethal because of inviability when tested as mature kernels, and we have attempted to rescue them by culturing their immature embyros. At the time the embryos were dissected from the immature kernels, they were examined to determine their stage of embryonic development.

Nutritional-type mutants are defined as mutants that are capable of germinating and growing into some form of a plant with both a shoot and roots when immature embryos are cultured or when tested at maturity.

Developmental-type mutants are those mutants that permanently are blocked in their development at some stage prior to the formation of leaf and root primordia (stage 1).

The two classes of mutants differ in their potential for the study of embryogenesis. On the one hand, the nutritional-type mutants, with their capability to produce a more or less normal-appearing plant with some form of a shoot and root (although they may be albino or otherwise abnormal in pigmentation), are thought to be defective in acquisition of nutrients or in some metabolic process. These mutants should be of value in examining the synthesis, uptake, or transport of nutritional substances.

The developmental-type mutants, on the other hand, should be of value in the study of the very early events of embryogenesis, including the control of the formation of the cell patterns that occurs during this period and that ultimately result in the development of leaf and root primordia and the shoot apical meristem [see Steeves and Sussex, 1972; Coe and Neuffer, 1978]. Since this group includes mutants that appear to be blocked at different points in the developmental sequence between the zygote and stage 1 (the presence of the first leaf primordium), they should aid in defining the precise steps that must occur in this sequence if normal development is to proceed.

A number of mutants that form one or more leaf primordia, have not yet been classified as nutritional-type mutants. Some of these have been tested and have failed to germinate when cultured. If these mutants continue to fail to germinate in culture, they will be considered a third type of mutant, one involving an abnormality in the regulation of dormancy. They may provide an opportunity to examine the control of the germination process.

A. Nutritional-Type Mutants

There are at least 108 nutritional-type mutants in this collection. These include 60 mutants known to be lethal because of inviability; ten mutants that are viable but display a mutant seedling phenotype; 33 mutants that are semi-lethal but are able to germinate and produce a weak, green plant at a low frequency; and, five mutants the status of which remains uncertain. It is among these 108 mutants that the search for auxotrophs appears most promising. The immature embryos of 85 of these mutants were cultured on basal and enriched medium [Sheridan and Neuffer, 1980]. Among these mutants, 16 displayed a superior shoot growth on the enriched medium, and the remainder grew about equally well on both kinds of media or, in a few cases, grew to a greater extent on the basal medium (Table I).

Among the 16 mutants showing promise for auxotrophy, one, mutant E1121, appears to be a proline-requiring mutant. This is evident from the observations that it is allelic to the *pro-1* mutant [Gavazzi et al, 1975; Racchi et al, 1978] and, that like the *pro-1* mutant, it is rescued when cultured on basal medium supplemented with proline (Fig. 4) [See Sheridan and Neuffer, 1980].

The observation that 85 of the mutants produced some form of a plant with a shoot (that in most cases was normally pigmented) and roots is remarkable in view of the fact that over two-thirds of these mutants are lethal and fail to germinate when tested as mature kernels. Yet, when cultured as immature embryos, they not only germinate but also are capable of producing some form of a complete plant. Although these mutants are generally retarded in their embryonic development and size when compared at the time of culturing to their normal counterparts on the same ear, it is, nevertheless, evident that their loss or severe impairment of viability must occur during the later stages of embryonic (and kernel) development.

The identification of 16 mutants that appear to favor a supplemented culture medium might be taken as encouraging results in our search for auxotrophic mutants of maize. Two features of these data suggest caution in their interpretation. First, it should be noted that in several cases, although the shoot growth displayed on the enriched medium was clearly superior to that obtained on the basal medium, nevertheless, the growth on the enriched medium was severely limited, resulting in very weak, small plants. Second, the single mutant, E1121, that has responded in a manner indicative of auxotrophy, is allelic to a previously reported mutant. Since the mutants included in this collection were selected from a much larger collection on the basis of severity of kernel defectiveness and

TABLE I
Promising Mutants With Greater Than 150% Shoot Growth on Enriched Medium Compared to Basal Medium*

	Enriched 1977				Enriched 1978				NH4-free enriched 1978		
E Number[a]	Weight E/B (%)	Mean weight E/B (mg)	Number of shoots E/B	E Number[a]	Weight E/B (%)	Mean weight E/B (mg)	Number of shoots E/B	E Number[a]	Weight E/B (%)	Mean weight E/B (mg)	Number of shoots E/B
744	317	073/023	05/05	873	850	051/006	10/10	931A	166	073/044	05/09
873	936	206/022	08/06	984A	197	059/030	08/11	1319A	189	278/147	18/09
1121	520	608/117	05/10	1024A	214	047/022	10/12	1404	198	315/159	07/08
1202A	227	148/065	07/07	1054	191	126/066	14/15				
1225B	1080	2.7/.25	08/06	1121	223	172/077	09/10				
1411	311	112/036	07/08	1255B	264	029/011	06/06				
1417	267	131/049	07/07	1308A	671	047/007	15/14				
				1373A	171	906/530	15/15				
				1392A	235	254/108	10/10				
				1405A	219	1166/533	10/10				
				1430	291	131/045	14/07				
				1431	182	178/098	15/15				
				pro-1	237	322/136	10/09				

*The number of plants harvested and used to determine the fresh weight of shoots is shown in the column labeled "Mean weight E/B," with the number to the right of the diagonal indicating the number for basal (B) medium and the number to the left of the diagonal indicating the number from the enriched (E) medium tested (enriched or NH4-free enriched). The ratio of these weights is presented in the "Weight E/B (%)" columns.

[a] All of the mutants are lethal and fail to germinate when tested as mature kernels except for 948A, which is uncertain, and E1431, which remains to be tested. The chromosome arm locations are E744, 91; E873, 9S; E948A, 1L; E1024A, 2L; E1121, 8; E1308A, 1S; E1417, 10L and pro-1, 8; while the other mutants, although tested, were not uncovered by the 18 B-A translocation stocks used in the test cross [Neuffer and Sheridan, 1980].

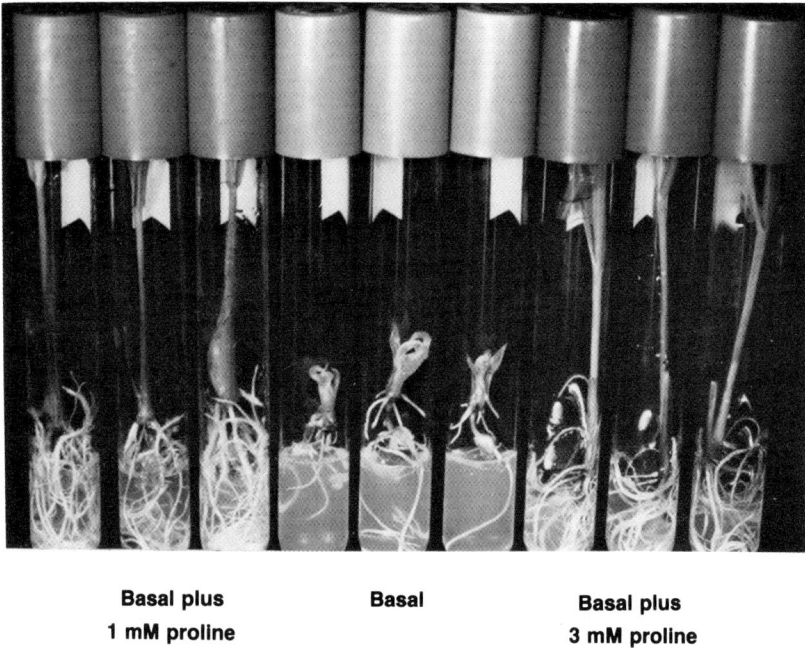

Basal plus	**Basal**	**Basal plus**
1 mM proline		**3 mM proline**

Fig. 4. Rescue of mutant, immature embryos of the *pro-1* mutant of Gavazzi et al [1975]. Immature mutant embryos were placed into culture 21 days after pollination and cultured for 21 days before photographing. The middle three tubes contained a basal medium of sucrose and salts; the left three tubes contained the basal medium supplemented with 1 mmole of proline; and, the right three tubes contained the basal medium supplemented with 3 mmoles of proline. Identical results were obtained with mutant E1121.

lethality in their original genetic background, with the goal of screening for potential auxotrophs, we now conclude that 1) the defective kernel type of mutants does not include even a moderate frequency of auxotrophs; or 2) our screening procedures are not revealing the auxotrophs in this collection, or 3) auxotrophs do not generally occur in maize. The only exception to the above conclusions appears to be the proline-requiring mutant, and its nature remains obscure.

The nutritional-type mutants have been tested for chromosome arm location, and the 56 mutants that have yielded at least one positive test are shown in Table II. The mutants have been located to 17 of the 18 arms that can be tested, with one to six mutants per chromosome arm. It is evident that the mutants of this type are located throughout the genome.

TABLE II
56 Nutritional-Type Mutants Located to Chromosome Arm*

1S	1L	2S	2L	3S	3L	4S	4L	5S	5L	6L	7L	8L	9S	9L	10S	10L
5	3	6	5	4	2	3	2	1	5	5	3	3	3	1	1	4
628A	948A	890A	660C	*874A*	*627D*	*211C*	*1005A*	*925A*	*1299*	874B	788	*1121*ᵃ	*873*	*744*	*1009*	*971*
923	*1303*	901A	*1024A*	1112	*936A*	*912*	*1156A*		*1331*	933	*1241 Pro-342*ᵃ		1089			*1330*
1142A		974A	1175	*1283*		931A			*1369*	*1176A*	*1332 Pro-1*		1421			*1417*
1208A		*1122A*	*1316A*	*1339A*					1308A	1294						*1427A*
1308A		1313	1414							1296A						
		1333B														

*Entries are included on the basis of one positive test. Those in italics have been confirmed by a subsequent test. No mutants were found on 6S; consequently, it is not listed in the table.

ᵃMutant 1121 and Coe's Pro-342 were not uncovered by any of the 18 arms tested but were located by two positive allelism tests with proline-1 located on chromosome 8 but not in the region uncovered by TB-8La [Gavazzi, et al, 1978].

B. Developmental-Type Mutants

Fifteen of the mutants in the total collection are tentatively classified as developmental-type mutants; all of these are lethal because of inviability, and they fail to form leaf primordia.

In addition to the 15 developmental-type mutants, 20 mutants are known to form at least one or more leaf primordium (stage 1 or later). Among these, eight have been tested and failed to germinate in culture, and the remaining 12 remain to be tested. These 35 mutants are listed according to stage of embryonic development in Table III. It should be noted that the ear age shown is for the oldest immature ear from which kernels were obtained for each mutant. Embryos from older ears, 25 or 30 days post-pollination, that are still blocked prior to leaf primordia formation (stage 1) are probably permanently blocked at the earlier stage.

Among the 15 developmental-type mutants, 11 are located to six chromosome arms (Table III). The nine mutants blocked at the proembryo stage of development occur at four, or possibly five, gene loci. They include mutant E792 and the four mutants allelic to it on chromosome arm 1S (E928A, E1348, E1394, and E1401) [Dooner, 1980] mutant E747B on chromosome arm 2S, mutant 1113A on chromosome arm 1L, mutant E1130 on chromosome arm 4L, and mutant E1425 (unlocated). The presence in our collection of five allelic mutants on the short arm of chromosome 1 is a consequence of the striking phenotype of this locus, which attracted our attention in selecting mutants for testing. Consequently, they are not a significant indication of the mutation frequency of this locus. Four mutants are blocked at the transition stage; mutants E1379A and E1428, both located on chromosome arm 2L, and E1311C and E1418 (unlocated). Two mutants are blocked at the coleoptilar stage, E1060B on chromosome arm 3L and E1399A (unlocated).

On this basis, it appears that at least six, and possibly as many as 11, separate mutants (at different gene loci) have been identified that block embryonic development at specific stages prior to formation of leaf primordia. All of the 15 mutants have been observed at a late enough stage in kernel development that it seems most likely that these mutants are truly blocked in their development. Two of these developmental-type mutants (E928A, and E1130), as well as three mutants that are slowed but not blocked in development (E330, E1092A, and E1309), are shown in Figure 5.

Among the remaining 27 mutants in our collection, 20 have been observed to reach stage 1 or later in development, and for seven mutants, the stage remains unknown. As shown in Table III, five mutants were observed at stage 1, five mutants were observed at stage 2, and ten mutants were observed at stage 3 or later. Together, these mutants represent at least seven different chromosome arms.

TABLE III
Developmental-Type Mutants and Other Mutants Classified According to Embryonic Stage at the Time of Observation*

Embryonic stage																	
Proembryo			Transition			Coleoptilar			Stage 1			Stage 2			Stage 3 or later		
Mutant E number	Ear age	Chromosome arm	Mutant E number	Ear age	Chromosome arm	Mutant E number	Ear age	Chromosome arm	Mutant E number	Ear age	Chromosome arm	Mutant E number	Ear age	Chromosome arm	Mutant E number	Ear age	Chromosome arm
747B	26	2S	1311C	40		1060B	39ª	3L	1092A	17		749	31		330D	35	3L
792	39ª	1S	1379A	36	2L	1399A	27		1119A	23		1172B	15		991	25	1S
928A	31	1S	1418	39					1147A	18		1196	15	5L	1104B	24	
1113A	32	1L	1428	30	2L				1191	20		1385	36	5L	1154A	26	
1130	25	4L							1315A	14	1L	1436A	22	2S	1309A	32	9L
1348	33	1S													1386A	20	
1394	26ª	1S													1390A	25	
1401	24	1S													1393A	39	
1425A	31														1410	35	
															1435	41	

*The classification of developmental stage of the embryo is based on that of Abbe and Stein [1954]; see Figure 1 for details. Normal embryos reached stage 1 by ten days, and kernels are fully mature by 50 days post-pollination in Missouri. The developmental-type mutants are the 16 mutants listed in the first three columns on the left; they appear to be blocked in the proembryo, transition, or coleoptilar stages of embryonic development.

ªObservation made in North Dakota, where generally cooler temperatures cause a somewhat slower development, stage 1 being reached by 16 days. Those seven mutants for which the embryonic stage remains undetermined are E1177A, E1210, E1253B, E1365, E1409, E1422, and E1429A.

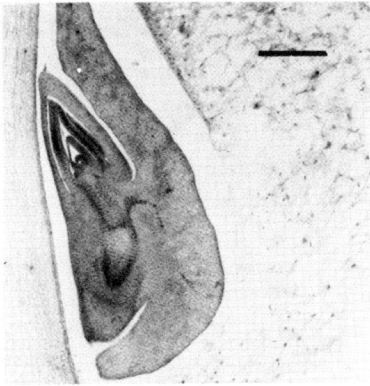

Normal St. 3 15 d.

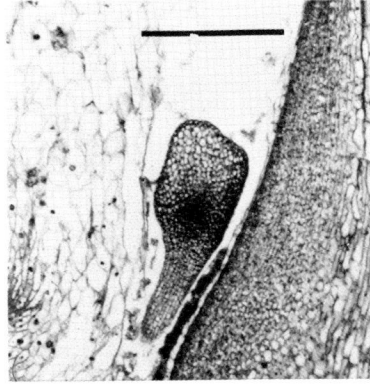

330 Trans. St. 3L 17 d.

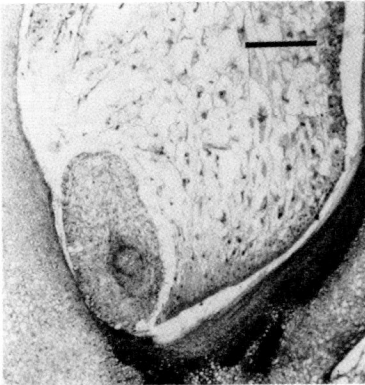

928A Proemb. St. 1S 15 d.

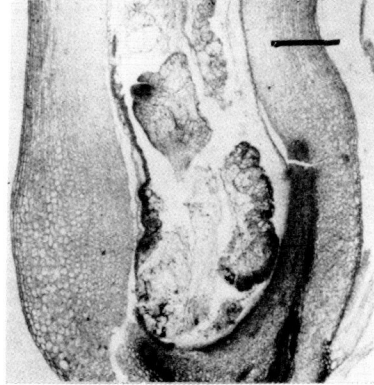

1092A Col. St. 1S 15 d.

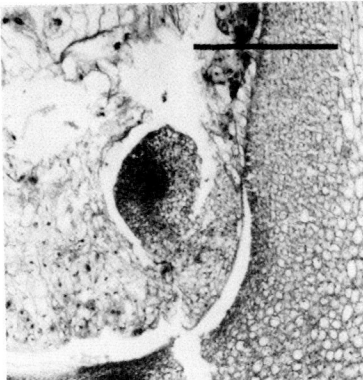

1130 Proemb. St. 4L 18 d.

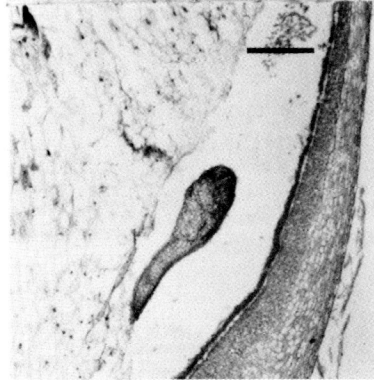

1309 Trans. St. 18 d.

Fig. 5. Longitudinal sections of normal and mutant, immature embryos. Mutant number, embryonic stage, chromosome arm location (except for 1309), and age in days post-pollination are shown. Note that they are as old as or older than the normal embryo (15 days and stage 3); but, in comparison, are blocked (E928A, and E1130) or retarded (E330, E1092A and E1309) in their development. The bar equals ½ mm in length.

VI.   EMBRYO–ENDOSPERM INTERACTION

All of the mutants described in this paper are defective in both their endosperm and embryo development. Consequently, it is evident that the expression of the mutant condition is not tissue specific.

In view of the oft-suggested role of the endosperm as a nurturing tissue during embryonic development (as well as at the time of germination), it may well be suspected that the absence of a normal endosperm in these mutants is the cause of the defectiveness of their embryos. Conceivably, then, the mutant genes of defective kernel mutants may be specific in their direct effect on the development of the endosperm and only secondarily, through this endosperm effect, bring about their defective embryonic development.

Our results from an examination of 19 mutants indicate that in most instances such is not the case. Rather, it appears that it is the genetic makeup of the embryo itself (or the endosperm itself) that determines its developmental fate. It has been possible by examining kernels on ears that revealed the chromosome arm location to determine 1) the developmental fate of a mutant embryo that develops in a kernel with a normal endosperm; 2) the fate of a normal embryo that develops in a kernel with a mutant endosperm, and 3) ways in which these two reciprocal alternatives compare with concordant mutant kernels (both tissues mutant) as well as completely normal kernels. We were fortunate in being able to carry out this analysis by using the B-A chromosome translocations. (See Neuffer and Sheridan [1980] for a more detailed description of the technique and results.)

When a mutant embryo developed in a kernel containing a normal endosperm, in 14 of the mutants there was no difference in the fate of the embryo compared to that in the concordant mutant kernel; in four mutants, the embryo was helped to some degree; and in one mutant, the results were inconclusive.

In the reciprocal arrangement, where a normal embryo developed in a kernel containing a mutant endosperm, in 15 cases the normal embryos grew into normal-appearing plants; in two cases they grew into weak plants; in one case the normal embryo grew into a tiny plant; and in the last case the genetically normal embryo was inviable. An examination of the degree of severity of mutant endosperm phenotype on these kernels as compared to that of concordant mutant kernels revealed that in three cases it was less extreme; in seven cases it was more extreme; and in nine cases there was no difference.

These results when examined in toto indicate that neither does a mutant endosperm usually harm the development of the embryo nor does the presence of a normal embryo usually aid the development of a mutant endosperm. Furthermore, a mutant embryo does not usually harm the developing endosperm to a large extent. Finally, and most significantly, usually a mutant embryo is not helped by the presence of a normal endosperm. Therefore, the developmental fates of the two tissues, endosperm and embryo, appear to be largely independent, and those cases of strong interaction or interdependency are likely to be the exception rather than the rule. It is evident that the notion that the embryo of a mutant kernel dies because the endosperm is defective is not true for a majority of the mutants analyzed. Rather, the genetic constitution of the embryo is what determines its fate in a majority of cases.

VII. DISCUSSION

A. Two Phases of Embryogenesis

The process of maize embryogenesis has two phases during the approximately 50-day period between pollination and kernel maturity. The first is the formation of the miniature plant from the zygote and its cellular progeny. This is achieved minimally with the formation of the first leaf primordium and root primordium, and normal embryos reach this stage by about ten days post-pollination under normal summer field conditions in Missouri. Such embryos can grow into normal-appearing plants when cultured on a simple medium (sucrose and mineral salts). Clearly, the essential events of embryo differentiation have occurred, and the embryo has achieved a developmental stage where it no longer has an absolute dependency on the maternal tissue for its growth into a plant.

The second phase is marked by the formation of additional leaf and, eventually, root primordia, and considerable enlargement of the scutellum and other embryo structures. This phase of addition and enlargement of the embryo is also characterized by its continued dormancy. The viviparous mutants [Robertson, 1955], although pleiotropic in their effect on aleurone and leaf pigment formation, are a class of mutants that escape from the dormancy normally maintained during this second phase, and consequently, precociously germinate on the immature ear.

B. Nature of Nutritional-Type Mutants

Those defective kernel mutants that are capable of germinating and growing into some form of a plant are obviously not impaired in the essential processes for primordia formation that occur during the first phase of embryogenesis. Although these mutants usually develop at a slower rate and are considerably reduced in size [Sheridan and Neuffer, 1980] compared to normal embryos, it is evident that they are capable of completing the first phase.

These mutants apparently are suffering from a defect in a metabolic or nutrient uptake process essential for the normal growth and enlargement that occurs during the second phase of embryogenesis. Some of these mutants may suffer from a defect in the chalazal cells, which normally serve as a conducting bridge between the ovary and basal cells of the endosperm, as reported for the viable defective kernel mutant *miniature (mn)* by Lowe and Nelson [1946]. A similar defect of normal kernel development was reported for the defective mutant *(de17)* by Brink and Cooper [1947]. They observed that the cells on the basal layer of the endosperm failed to differentiate into an absorbing tissue, resulting in a limitation of endosperm and embryo growth; however, the embryos were often viable and grew into plants. That many of the 108 nutritional-type mutants in our study are of this type is probable, inasmuch as over half of the 85 mutants tested grew about equally well on both basal and enriched medium and 32 of the 108 mutants are viable.

Nevertheless, many of these mutants must suffer from a defect in some process other than nutrient uptake. This is indicated by the variety of altered phenotypes such as dwarf, striped, narrow leaves, lack of pigments, weak spindly plants, and other abnormalities observed upon germination of cultured immature embryos. (See Sheridan and Neuffer [1980] for a more detailed description.) This conclusion is also indicated by the superior shoot growth rate on enriched medium exhibited by some of the mutants.

C. Nature of the Developmental-Type Mutants

The 15 mutants representing at least six gene loci that were observed to be at the proembryo, transition, or coleoptilar stage of embryogenesis (at a time when normal embryos on the same ear, were at stage 1, or usually at a much later stage) may suffer their defect in some process of the first phase of embryogenesis.

The nature of the defect in these mutants remains unknown. That they do not suffer a total block in cell division, protein synthesis, or some other essential cellular process is indicated by their capacity for considerable growth beyond the zygote. Similarly, it does not seem likely that they suffer a block in nutrient uptake or some metabolic synthesis since mutants of that sort, to the limited degree that they are known, appear capable of proceeding through the first phase of embryogenesis.

Rather, it seems likely that these mutants suffer from a defect in their determination or differentiation processes so that the first phase of embryogenesis is interrupted. The nature of these processes is essentially unknown for plant embryogenesis; however, some of them may involve hormonal mechanisms for the regulation of development. In any case, it is essential that a detailed view of the developmental fate, throughout the duration of kernel development, be obtained for each of these mutants in order to more fully understand the process they block or disrupt.

Among the 20 mutants that formed at least one leaf primordium, eight were cultured as immature embryos that had reached stage 1 or later in development, and all failed to germinate. These eight mutants are apparently incapable of germinating in culture as well as at maturity; whether the other 12 mutants also lack this capacity remains to be determined by culture tests. Similarly, the seven mutants for which embryonic stage is not yet determined may be of this type.

The reasons that some of the mutants possessing primordia fail to germinate precociously when cultured (as normal embryos do) remain unknown. The presence of the primordia apparently rules out a block in the differentiation process. It is probable that some of them are really nutritional-type mutants and, indeed, some of these mutants probably will be moved into that group. It seems likely, however, that some of these mutants fail to germinate because of a defect in their control of dormancy. Since this failure is exhibited by isolated embryos separated from the endosperm, the defectiveness of the mutant embryos must result from the mutant genotype in the embryonic tissues rather than the mutant condition of the endosperm. Although this same specificity of genetic control is present in the viviparous mutants (ie, the genotype of the embryo is determinative for vivipary [Robertson, 1955]), we suspect that the defect in the dormancy mechanism in these mutants may be the reverse of that in the viviparous mutants. Rather than a premature loss of the dormancy mechanism (abscisic acid level or sensitivity), these mutants may be permanently dormant, at least under usual test conditions.

From the results to date, it is nevertheless evident that several gene loci have mutated to block primordia formation, and that an even greater number of loci have mutated to severely impair this process. It is not known what proportion of the former loci are the same as the many loci that have mutated to produce the nutritional-type mutants.

X. SUMMARY

A collection of 150 defective kernel maize mutants have been examined by genetic, morphological, and embryo culture techniques. All of the mutants are defective both in endosperm and in embryo development, and all are single gene, recessive mutants; the majority are lethal.

Because all of the mutants display their defective endosperm phenotype at an early age on immature self-pollinated ears, the mutant kernels and their embryos have been examined, and 108 mutants have been cultured. On the basis of these tests and other observations, we have classified most of the mutants into two groups according to operational definitions.

The 108 nutritional-type mutants are those mutants that are capable of germinating and growing into some form of a plant with both a shoot and roots when immature embryos are cultured or when tested at maturity. Fifty-six of these mutants have been located to 17 different chromosome arms. Sixteen show some promise in our search for auxotrophy, and one of these, E1121, is a proline-requiring mutant that is allelic to the *pro-1* mutant of Gavazzi et al [1975].

The 15 developmental-type mutants are those mutants that are blocked in their development prior to the formation of leaf and root primordia; 11 of these mutants have been located to six different chromosome arms.

The nutritional-type mutants are believed to be defective in acquisition of nutrients or some metabolic process. The developmental-type mutants are believed to be altered in the first phase of embryogenesis, prior to or during primordia formation.

ACKNOWLEDGMENTS

This research was supported by grants from the National Science Foundation (PCM76-10562, PCM78-08559, and PCM 80-05709 to M. G. Neuffer and PCM76-10563, PCM78-08560, and PCM80-04814 to W. F. Sheridan).

REFERENCES

Abbe EC, Stein OL (1954). Am J Bot 41:285–293.
Baker WK (1978). Annu Rev Genet 12:451–470.
Beckett JB (1978). J Hered 69:27–36.
Brink RA (1927). Am Nat 61:520–530.
Brink RA, Cooper DC (1947). Genetics 32:350–368.
Coe EH Jr, Neuffer MG (1978). In "The Clonal Basis of Development." (S Subtelny and IM Sussex, eds.) Academic Press, New York.
Demerec M (1923). J. Hered 14:297–300.
Dooner HK (1980). Maize Genet Coop News Letter 54:79–80.
Emerson RA (1932). Science 75:566.
Gavazzi G, Nava-Rachi M, Tonelli C (1975). J Theoret Appl Genet 46:339–346.
Gavazzi G, Todesco G, Neuffer MG (1978). Maize Genet Coop Newsletter 52:66–67.
Gehring WJ (1976). Annu Rev Genet 10:209–252.
Kiesselbach TA (1949). University of Nebraska Agr Exp Sta Bull 161.
Lowe J, Nelson OE Jr (1946). Genetics 31:525–533.
McLaren A (1976). Annu Rev Genet 10:361–388.
Mangelsdorf PC (1926). Conn Agr Exp Sta Bull 279:509–614.
Meinke DW, Sussex IM (1979a). Dev Biol 72:50–61.
Meinke DW, Sussex IM (1979b). Dev Biol 72:62–72.
Murashige T, Skoog F (1962). Physiol Plant 15:473–497.
Neuffer MG, Sheridan WF (1980). Genetics 95:929–944.
Paxson JB (1963). Some effects of plant regulators on embryogeny in Zea mays. Ph.D. dissertation. Texas A & M University.
Racchi ML, Gavazzi G, Monti D, Manitto P (1978). Plant Sci Letts 13:357–364.
Randolph LF (1936). J Agr Res 53:881–916.
Robertson DS (1955). Genetics 40:745–760.
Roman H, Ullstrup AJ (1951). Agron J 43:450–454.
Sass JE (1968). "Botanical Microtechnique." Iowa State College Press, Ames, Iowa.
Sheridan WF, Neuffer MG (1980). Genetics 95:945–960.
Steeves TA, Sussex IM (1972). In "Patterns in Plant Development." pp. 27-29. Prentice-Hall, Inc., Englewood Cliffs, New Jersey.
Wang FH (1947). Am J Bot 34:113–125.
Wentz JB (1925). Genetics 10:395–401.
Wentz (1930). Iowa Ag Exp Sta Res Bull 121:345–379.

Globin Gene Expression in Chick Embryogenesis

Robert W. Keane and Vernon M. Ingram

Department of Biology, Massachusetts Institute of Technology, Cambridge,
Massachusetts 02139

I. INTRODUCTION

Biochemical investigations of determination and differentiation have been hampered by the lack of guidance from convincing models of these processes at a molecular level. Therefore most biochemical and molecular research in this area has of necessity been focused on the expression, and the control of expression, of gene products characteristic of various differentiated tissue types. Such studies are most often concerned with differentiation, a late stage of development. Determination is not directly investigated, although the information that results from studies of differentiation will affect the formulation of models of determination.

An example of deriving, perhaps prematurely, a hypothetical molecular mechanism for differentiation from structural biochemical information is found in

Levels of Genetic Control in Development, pages 157–169

recent work on the location and order of embryonic and adult globin genes in humans [Fritsch et al, 1980]. Through recombinant DNA technology the genes coding for the various embryonic, fetal, and adult human "β-like" globin genes have been located on the chromosomal DNA and placed in order next to each other. Remarkably the order in which the β-globins appear during development— embryonic, fetal, adult—is also the order in which they are located on the DNA. This is also the direction of transcription of the individual genes. One is tempted to assume that the geographical sequence controls the temporal sequence, with perhaps transcription as the mechanism. This interpretation is, however, almost certainly too facile, since the structural genes are separated by long stretches of DNA with no known function, and since there must also be a temporal sequence for turning off embryonic and fetal globin genes. There might be "control" sequences in the in-between stretches of DNA, by analogy with pro-karyotic cells, and these controls might be operated by some soluble factor. Such a detailed structural view of a differentiating system provides the factual foundation in which a mechanistic model of differentiation of that cell type will operate.

The available experimental evidence argues that most or nearly all of the cells of any one organism contain the same genome [Briggs and King, 1952; Gurdon, 1964]. It is generally assumed that genes coding for the structure of specific differentiated products in different adult cell types exist in the same form in the germ line DNA and also in the DNA of the embryo's cells. The events that lead to the expression of a specific gene are viewed as operating on a given DNA sequence, causing it to be expressed rather than due to changes in the DNA sequence itself. Specific changes in DNA methylation have been reported to occur during development [McGhee and Ginder, 1979]. These changes are characteristic for a given cell type and presumably are unique for the cell lineage. In the case of the globin gene region there appears to be a loss of methyl groups in DNA of precursors for the red blood cells during terminal stages of differ-entiation [van der Ploeg and Flavell, 1980]. However, the relationship between DNA methylation and gene activity is still unclear.

We do know which cells are most useful for studies of differentiation with all the biochemical techniques available. We select our system to have a sufficient supply of such cells, hoping we will be able to abandon the "shotgun" bio-chemical approach in favor of a directed, more incisive attack.

There is at present no model to guide us in investigations of cell determination. We do not know whether to look at specific gene products or if so, what products might be important. We cannot distinguish cells before, during, or after their determination, and the small numbers present during this process make bio-chemical analysis almost impossible. Yet determination is one of the most im-portant events in development. In this paper we describe a novel approach to the problem that involves arresting development through virus transformation and using the transformed state to produce a clone from a single precursor, thus providing sufficient material for biochemical work. While the objective of transforming predetermined, multipotential cells has not yet been achieved, we have obtained promising results with transformed derivatives of chick embryonic erythroid precursors.

II.   MODEL SYSTEMS TO STUDY DIFFERENTIATION

A. Mouse Teratocarcinoma Cells

Mouse teratocarcinoma cells have served as a model system for the study of cell differentiation. Teratocarcinomas are malignant tumors characterized by the presence of a distinctive cell type known as embryonal carcinoma cells. Teratocarcinomas contain other differentiated cell types, such as nerve, cartilage, muscle, and skin. Embryonal carcinoma cells are stem cells of these tumors and show remarkable similarities to the cells of the early embryo. Embryonal carcinoma cells are totipotent and can be injected into normal embryos at the blastocyst stage, where they differentiate into all somatic and germinal tissues [Brinster, 1974; Mintz and Illmensee, 1975; Illmensee and Mintz, 1976]. The two cell types share at least one common antigen, and teratocarcinomas can be obtained by implantation to early embryos in extrauterine sites [Artzt et al, 1973].

Several clonal cell lines of embryonal carcinoma cells have been established [Finch and Ephrussi, 1967; Kahan and Ephrussi, 1970; Rosenthal et al, 1970]. Differentiation of embryonal carcinoma cells can occur in culture and therefore provides a means for studying cell determination and differentiation. Embryonal carcinoma cells lose their malignancy when they differentiate in vivo, suggesting that teratocarcinoma cells may be useful for studying the relation between neoplasia and differentiation. The malignancy of embryonal carcinoma cells may be a function of their similarity to early embryo cells, and the transition from normal pluripotent embryonic cells to malignant embryonal carcinoma cells is readily reversible.

Markert [1978] has suggested that there is an inverse relation between normal control mechanisms operating in cell differentiation and those governing transformation. The teratocarcinoma cell system has most elegantly demonstrated that a malignant cell can make a transition to a normal pluripotent cell when injected into the blastocyst embryo. Embryonic cells and embryonal carcinoma cells are also similar in that they are nonpermissive for tumor virus replication [Jaenisch and Berns, 1977], but they can become permissive for tumor virus during subsequent differentiation.

B. Murine Erythroleukemia Cells

The biochemistry of cells during terminal differentiation has been investigated in a number of developing systems. Murine erythroleukemia cells have been most vigorously investigated and have contributed much to our understanding of erythroid development [Marks and Rifkind, 1978]. Murine erythroleukemia cell lines have been produced by transformation of fetal spleen cells with the Friend virus complex, which includes at least two viruses, a defective spleen-focus-forming virus and a murine leukemia helper virus.

Some murine erythroleukemia cell lines show a low level of spontaneous differentiation; they can be induced to resume erythroid differentiation by culture with a wide variety of agents [Marks and Rifkind, 1978]. This observation suggests that there may be different pathways by which such agents induce differentiation. Induction of murine erythroleukemia cell differentiation usually follows a latent period, and there appears to be a requirement for a cell-cycle-related event for murine erythroleukemia cells to become committed [Marks and Rifkind, 1978]. Once the program of differentiation has been initiated, however, it shows many similarities to normal erythropoietin-regulated differentiation.

III. TRANSFORMATION AND CELL DIFFERENTIATION

Electron microscopic observation of virus-like structures in embryos and the detection of virus-related proteins by immunologic techniques have generated the idea that RNA tumor viruses may be important for embryonic development and have a function in cell differentiation [Jaenisch and Berns, 1977].

Transforming viruses have recently been used to interfere with differentiation by blocking the development of precursor cells at some precise point. Some, but not all, of the differentiated functions may be altered and may be useful in determining normal pathways of development. This blockade of development has been reported for Rous sarcoma virus (RSV)-transformed myoblasts [Fiszman and Fuchs, 1975; Fiszman, 1978], chondroblasts [Muto et al, 1977; Pacifici et al, 1977], retinal melanocytes [Boettiger et al, 1977], neural retinal cells [Pessac and Calothy, 1974], wild-type *(wt)*-avian erythroblastosis virus (AEV) and temperature-sensitive *(ts)*34-AEV-transformed erythroblasts from adult bone marrow [Graf et al, 1978a, 1978b], and SV40-transformed teratocarcinoma cells [Topp et al, 1978].

Chick embryo myoblasts and retinal melanoblasts have also been transformed by a temperature-sensitive mutant of Rous sarcoma virus (RSV-ts68). The transformed phenotype can be reversed and the normal program of differentiation partially restored by a temperature shift from 37°C (permissive temperature) to 41°C (nonpermissive temperature).

IV. A NEW APPROACH TO BIOCHEMICAL INVESTIGATIONS OF EARLY EMBRYOGENESIS

We have developed a new approach to the biochemical investigation of determination and differentiation that will also give insights into the relation between cell differentiation and neoplasia. Techniques have been developed for the isolation of populations of undifferentiated primary mesenchymal cells from the early chick embryo, free of epiblast (ectoderm) and hypoblast (endoderm) cells. These mesenchymal cells have been infected by a transforming virus to stop their differentiation. The transformed derivatives were cloned and grown to large populations. Both wild-type virus and a mutant, temperature sensitive for transformation, were used. By chemical means or temperature shift experiments, the cells were stimulated to express their differentiation (at least partially) so that these processes could be examined biochemically.

The primary mesenchyme is a collection of cells that gives rise to a number of differentiated cell types—for example, blood, muscle, bone, and connective tissue. We have concentrated on the development of erythroid cells. The uninfected early embryo elaborates a nonsustaining primitive red cell population, which is replaced by a self-sustaining definitive red cell population. Each cell line synthesizes different hemoglobins and polypeptide chains (Fig. 1). It is not clear whether the primitive and definitive cell series share a common stem-cell population or whether they are each derived from different stem cells.

A. Mesenchymal Cell Cultures

Figure 2a shows techniques for the isolation of populations of undifferentiated primary mesenchymal cells from definitive streak or head process stage chick embryos. Blastoderms were removed from the egg and transferred to a dish containing sterile Howard's Ringers solution. The posterior two-thirds of the area pellucida and definitive streak were cut out and the resulting tissue pieces were incubated in Ca^{++},Mg^{++}-free Tyrode's solution (CMF) containing 0.25% collagenase and 0.25% pancreatin for 12 minutes at 37°C. The tissue pieces were then pulled apart with watchmaker's forceps to yield a population of mesenchymal cells free of epiblast and hypoblast.

Figure 2b shows a scanning electron micrograph of the three primary germ layers of the chick embryo after collagenase-pancreatin digestion. Epiblast and hypoblast separate cleanly from the mesoblast leaving a sheet of pure mesenchymal cells. All cells of the isolated primary mesenchyme attach to the culture dish five hours after plating and form a sheet of cells. After 24 hours of culture of the isolated mesoblast, various cell types are clearly recognizable: for example, beating heart muscle, erythroid cells, epithelial cells, and fibroblastic cells.

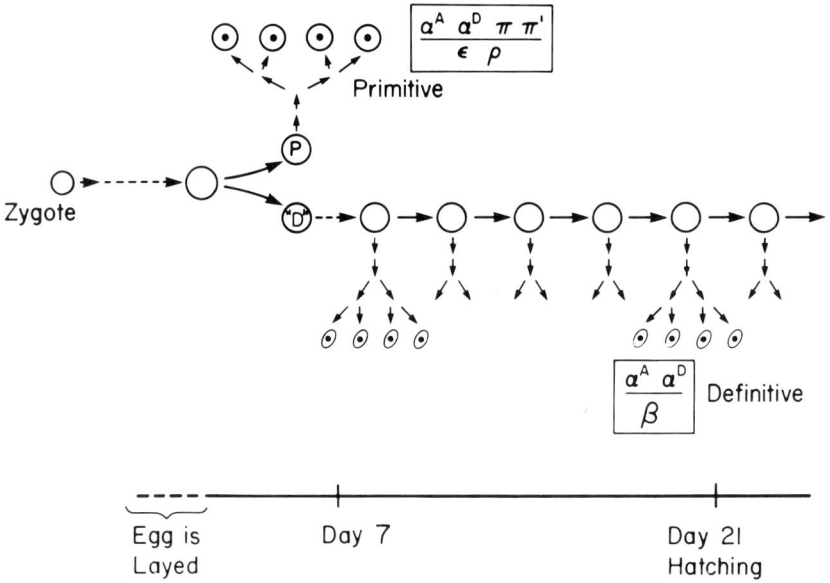

Fig.1. Schematic time course of the successive appearance of primitive and definitive cells during development of the chick embryo. Polypeptide chains of hemoglobins for the two cell types are divided into the α-like chains above the line and the β-like chains below the line. The change from primitive to definitive cell production in the chick embryo occurs about the fifth day of incubation [chick nomenclature from Brown and Ingram, 1974]. Reproduced from Annals of Internal Medicine (1980) 92:547–552 (© 1980 Annals of Internal Medicine; used with permission).

B. Erythroid Cells and Globin Polypeptide Chains Synthesized in Mesenchymal Cell Cultures

Most of the attached cells that form in mesenchymal cell cultures are erythroid cells. Clusters of cells with erythroid morphology adhere to the culture dish at 24 hours of culture, but float freely in the culture medium at 48 hours. These cells are oval or round with a round condensed nucleus and are similar in structure to mature primitive and definitive cells in ovo [Bruns and Ingram, 1973]. The globin chains synthesized by these erythroid cells are shown in Figure 3. At four hours after isolation, mesenchymal cells synthesize a major globin band that co-migrates with the π-chain standard and some minor bands in the β-region of the gel. By 24 hours of culture, mesenchymal cells are clearly producing "α-like" chains (π, α^A and α^D) and a small amount of ε chains ("β-like"). An increase in the proportion of the β-like chains (ε, ρ, β) was observed between 24 and 48 hours of culture, and the amounts of both the α- and β-like chains produced

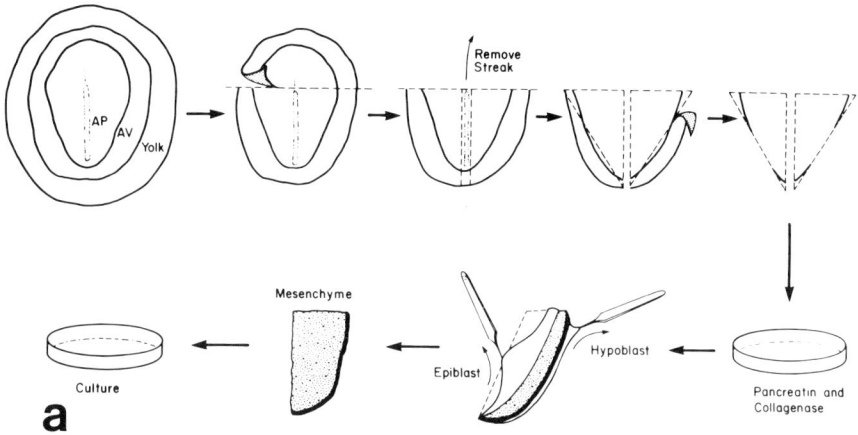

Fig. 2a   Procedure for isolation of mesenchymal cells, area pellucida (AP), area vasculosa (AV). Reproduced from Cell (1979) 17:801–811 (© 1979 Massachusetts Institute of Technology; used with permission).

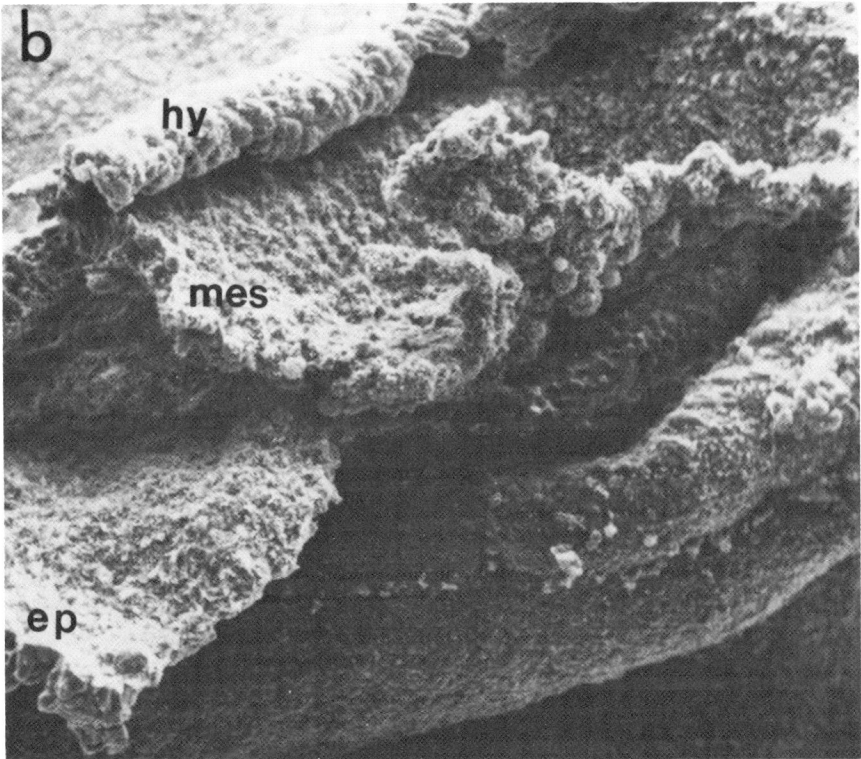

Fig. 2b.   Separation of the three primary germ layers of chick embryo, stage 6. The epiblast (ep) and hypoblast (hy) are rolled back to free the mesoblast (mes) as a single sheet of cells. Reproduced from Cell (1979) 17:801–811 (© 1979 Massachusetts Institute of Technology; used with permission).

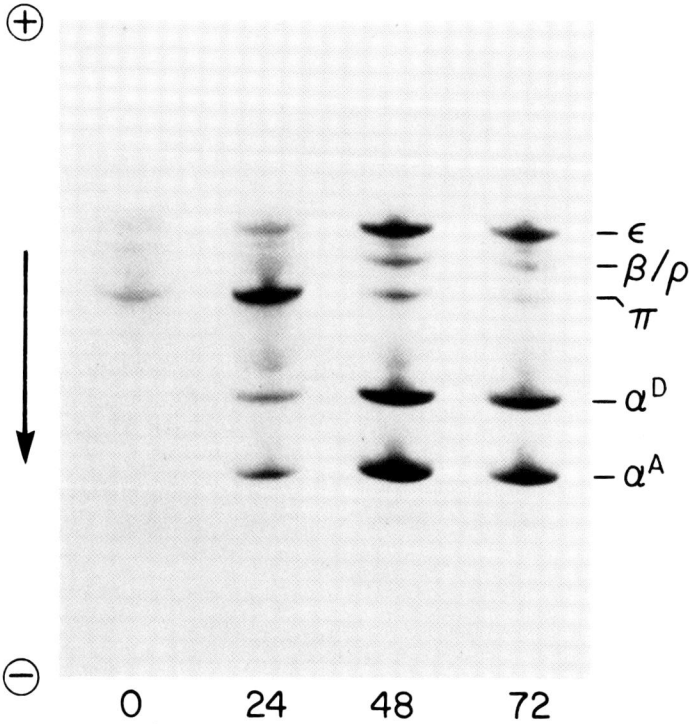

Fig. 3. Globin chains synthesized in mesenchymal cell cultures at 0, 24, 48, and 72 hours of culture. Mesenchymal cultures were labeled with ^{14}C-leucine-lysine for 4, 24, 48, and 72 hours. Cells were washed, lysed, and immunoprecipitated with excess anti-adult chicken hemoglobin. Immunoprecipitates were run on acid urea polyacrylamide gels containing 0.6% Triton X-100 (Sigma Chemical Co., St. Louis, MO) and then fluorographed. Reproduced from Cell (1979) 17:801–811 (© 1979 Massachusetts Institute of Technology; used with permission).

did not change on further cell culture. Thus it appears that the α-like globin chains (π, α^A, α^D) are expressed nearly 24 hours before the β-like globin chains (ϵ, ρ, β) can be detected.

C. Transformation of Mesenchymal Cells With *wt*-AEV and *ts*34-AEV

The cloned transformants of *wt*-AEV or *ts*34-AEV examined to date have not shown the presence of more than one cell type within a clone, nor the presence of two or more different characteristic differentiation products in individual cells. The question of whether either situation could exist remains

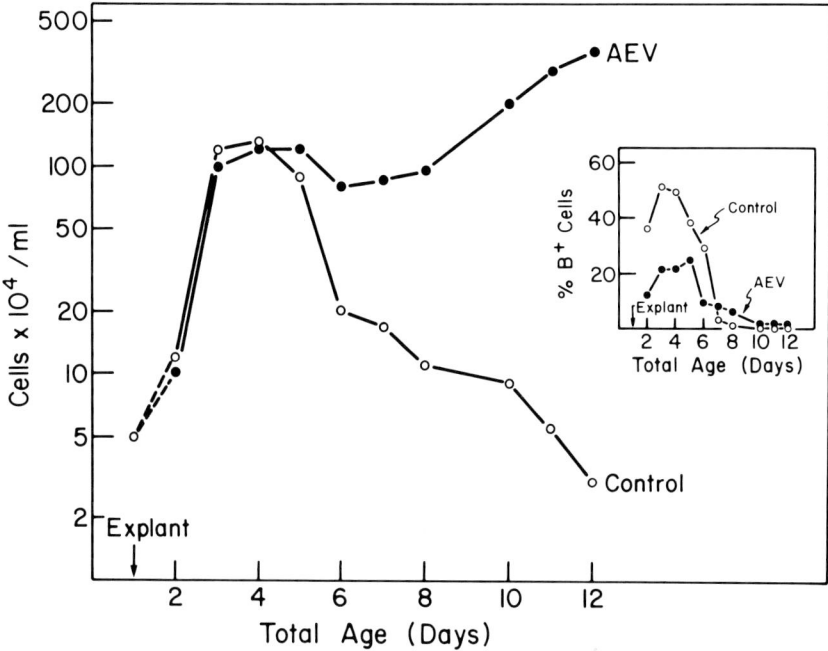

Fig. 4.   Total number of suspended cells in uninfected (○—○) and wild-type avian erythroblastosis virus (AEV)-infected (●—●) mesenchymal cell cultures. Percentages of suspended cells that stain with benzidine are shown in the inset. Reproduced from Cell (1979) 17:801–811 (© 1979 Massachusetts Institute of Technology; used with permission).

unanswered, because the clones examined have not been tested with the appropriate multiple differentiation probes and because only a few clones have thus far been examined. Transformants of primary mesenchyme with AEV have been selected for their ability to grow and clone in suspension culture, favoring the selection of clones with erythroid properties.

In addition to causing cell proliferation (we have selected for this property), wt-AEV (and ts34-AEV) blocks the expression of specific genes responsible for the production of differentiated proteins (Fig. 4). Thus far we have studied this phenomenon only in the erythroid series with respect to specific globin chain production, heme production, and erythrocyte-specific histone H5.

Clones of wt-AEV-transformed mesenchymal cells (MAE cells) express α^D and a small amount of π chains, but no normal α^A- or β-type globin chains (β, ϵ, ρ) are produced (Fig. 5). The ts34-AEV-transformed mesenchymal cells (MAE-ts34 cells) express both α^D and α^A, but again no recognizable β-type

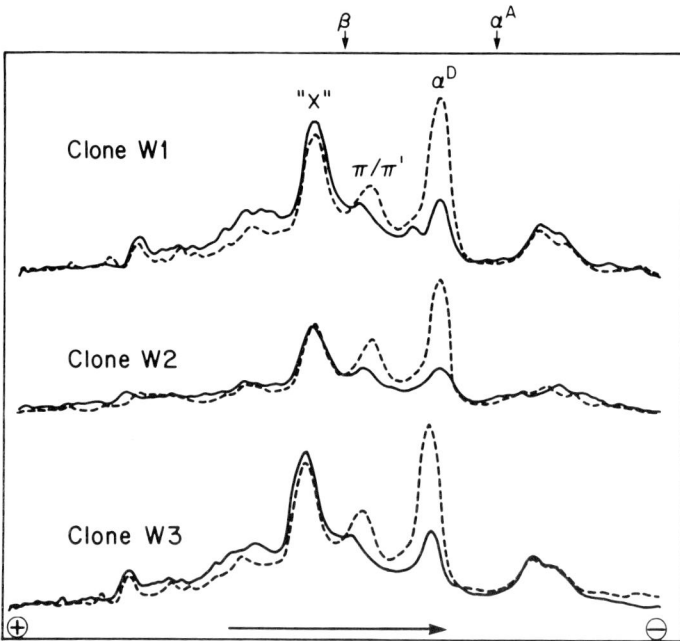

Fig. 5. Representations of fluorographic scans of immunoprecipitated ^{14}C-leucine/lysine-labeled globins of wt-AEV-transformed mesenchymal cell (MAE cell) clones W1, W2, and W3 untreated (—) or treated with 1.0 mM n-butyrate (- - -). Reproduced from Cell (1979) 17:801–811 (© 1979 Massachusetts Institute of Technology; used with permission).

chains are produced (Fig. 6). True β-type globins are therefore absent from clones derived from both wt-AEV and ts34-AEV infections. We might be dealing solely with transformed primitive erythroid cells. Possibly the two viruses restrict the expression of globin chain production in a different manner or transform target cells at different points in the erythrocyte developmental pathway. The ts-AEV may interrupt the sequence of appearance of globin chains later than wt-AEV, after α^A globin chain synthesis has been initiated but before β (or ε or ρ) appears.

A number of other proteins in transformed cells are precipitated by the antibody against whole hemoglobin. These proteins have been shown to be non-specific cellular proteins, immunoprecipitated by the antibody against whole adult chicken hemoglobin.

Transformation of primary mesenchyme with AEV enabled us to separate the control of heme from globin synthesis. Wild-type AEV suppresses heme synthesis, but the level of inhibition varies with different clones. There does not seem to be any quantitative correlation between the degree of inhibition of heme

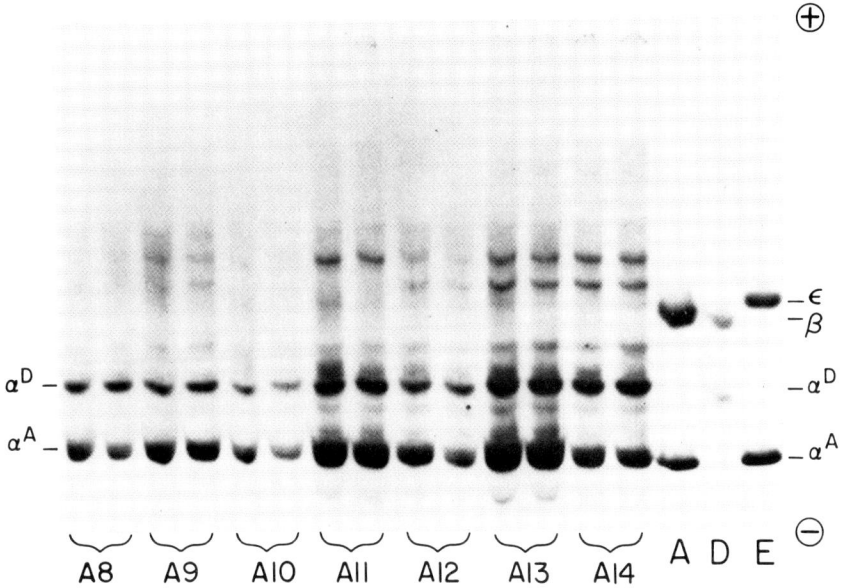

Fig. 6.   Fluorograph of globin of *ts*34-AEV-transformed mesenchymal cell (MAE-*ts*34 cell) clones immunoprecipitated with anti-adult chicken hemoglobin and separated by electrophoresis in acid urea polyacrylamide gels containing 0.6% Triton X-100. For each MAE-*ts*34 clone (A8 through A14), the first lane represents immunoprecipitate of cells at 37°C, and the second lane represents immunoprecipitate at 41°C. The ^{14}C-leucine/lysine-labeled hemoglobin standards A and D were from 12-day embryonic red cells; hemoglobin E was from four-day embryonic erythroid cells. Reproduced from Cell (1979) 17:801–811 (© 1979 Massachusetts Institute of Technology; used with permission).

synthesis and the degree of inhibition of globin synthesis. Further, MAE-*ts*34 cells show temperature-sensitive inhibition of heme synthesis, as shown in Table I. This promises to be an interesting set of transformants, because temperature-sensitive inhibition of heme synthesis should enable us to ascertain the mechanism of this inhibition in future experiments. In addition, information concerning the normal developmental processes involved in the onset of heme synthesis may result. The chick erythrocyte-specific histone H5 is also under temperature-sensitive control in MAE-*ts*34 cells. At the lower temperature (permissive for transformation), some clones have a striking inhibition of H5 synthesis, which is partially restored at the higher temperature (nonpermissive for transformation). Use of this phenomenon will throw light on the control of H5 synthesis.

Clones of MAE cells make only the α^D-globin chain and a small amount of the π-globin chain. The addition of 1 mM of n-butyrate, a known inducer of differentiation in murine erythroleukemia cells, causes a dramatic increase in

TABLE I
Percentages of Benzidine-Positive *ts*34-AEV-Transformed Mesenchymal Cells (MAE-*ts*34 Cells), Hemoglobin-Positive MAE-*ts*34 Cells, and Chicken Histone H5-Positive MAE-*ts*34 Cells*

Clone	Temperature (°C)	Shift 1 (%)	Shift 2 (%) Day 0[a]	Day 1	Day 2	Day 3
Benzidine-positive MAE-*ts*34 cells						
A1	37	4	5	3	2	2
	41	64		24	25	26
A2	37	20	24	25	25	20
	41	79		30	28	27
A3	37	10	9	4	4	4
	41	47		12	8	8
Hemoglobin-positive MAE-*ts*34 cells						
A1	37		ND	4	4	14
	41			27	28	27
A2	37		ND	20	28	19
	41			33	28	27
A3	37		ND	7	12	18
	41			15	18	30
Chicken histone H5-positive MAE-*ts*34 cells						
A1	37		ND	3	5	13
	41			20	24	27
A2	37		ND	2	10	14
	41			10	14	27
A3	37		ND	13	10	15
	41			31	45	30

*Reprinted from Cell (1979) 17:801–811(© 1979 Massachusetts Institute of Technology; used with permission).
[a]ND = not done.

the production of the α^D-globin chains as well as an increase in the production of π-globin chains (see Fig. 5). The unidentified protein (x) is not affected by this treatment. This observation is reminiscent of the effect of n-butyrate in causing a very large increase in the production of certain specific hormones such as human chorionic gonadotropin [Lieblich et al, 1976]. There may be a similar mechanism involved in the induction by n-butyrate of differentiation-specific globins and hormones in murine erythroleukemia cells, HeLa cells, and MAE cells.

V. CONCLUSIONS

This paper demonstrates the usefulness of viral transformation of early embryonic mesenchymal cells in studies of determination and differentiation. The large numbers of homogeneous tranformed cells in a given state of differentiation obtained permit a variety of hitherto unapproachable molecular studies. Further exploitation of these techniques will permit the examination of properties of malignant cells as well as the transformation of normal cells in culture.

REFERENCES

Artzt K, DuBois P, Bennett D, Condamine H, Babinet C, Jacob F (1973). Proc Natl Acad Sci USA 70:2988–2992.
Boettiger D, Roby K, Brumbaugh J, Biehl J, Holtzer H (1977). Cell 11:881–890.
Briggs R, King TJ (1952). Proc Natl Acad Sci USA 38:455–463.
Brinster RL (1974). J Exp Med 140:1049–1056.
Brown JL, Ingram VM (1974). J Biol Chem 249:3960–3972.
Bruns GAP, Ingram VM (1973). Phil Trans R Soc Lond Biol 266:225–305.
Finch BW, Ephrussi B (1967). Proc Natl Acad Sci USA 57:615–621.
Fiszman MY (1978). Cell Diff 7:89–101.
Fiszman MY, Fuchs P (1975). Nature 254:429–431.
Fritsch EF, Lawn RM, Maniatis T (1980). Cell 19:959–972.
Graf T, Ade N, Beug H (1978). Nature 275:496–501.
Graf T, Beug H, Royer-Pokora B, Meyer-Glauner W (1978b). In "Differentiation of Normal and Neoplastic Hematopoietic Cells." (B Clarkson, PA Marks, and J Till, eds.), pp 625–639. Cold Spring Harbor Laboratory, Cold Spring Harbor, New York.
Gurdon JB (1964). In "Advances in Morphogenesis." (M Abercrombie and J Brachet, eds.) vol 4, pp 1–43. Academic Press, New York.
Illmensee K, Mintz B (1976). Proc Natl Acad Sci USA 73:549–553.
Jaenisch R, Berns A (1977). In "Concepts in Mammalian Embryogenesis." (MI Sherman, ed.), pp 287–314. M.I.T. Press, Cambridge, Massachusetts.
Kahan BW, Ephrussi B (1970). J Natl Cancer Inst 44:1015–1036.
Lieblich JM, Weintraub BD, Rosen SW, Chou JY, Robinson JC (1976). Nature 260:530–532.
Markert CL (1978). In "Cell Differentiation and Neoplasia." (MI Sherman, ed.), pp 287–314. M.I.T. Press, Cambridge, Massachusetts.
Marks PA, Rifkind RA (1978). Annu Rev Biochem 47:419–448.
MeGhee JD, Ginder GD (1979). Nature 280:419–420.
Mintz B, Illmensee K (1975). Proc Natl Acad Sci USA 72:3583–3589.
Muto M, Yoshimura M, Okayama J, Kaji A (1977). Proc Natl Acad Sci USA 74:4173–4177.
Pacifici M, Boettiger D, Roby K, Holtzer H (1977). Cell 11:891–899.
Pessac B, Calothy G (1974). Science 185:709–710.
Rosenthal MD, Wishnow RM, Sato GH (1970). J Natl Cancer Inst 44:1001–1014.
Topp W, Hall JD, Rifkin D, Levine AJ, Pollack R (1978). J Cell Physiol 93:269–276.
van der Ploeg LHT, Flavell RA (1980). Cell 19:947–958.

Do Functional Properties of the Cell Membrane Play a Crucial Role in Murine Erythroleukemia Cell Differentiation?

Robert Levenson, Lewis Cantley, and David Housman

Center for Cancer Research and Department of Biology, Massachusetts Institute of Technology (R.L., D.H.), and Department of Biochemistry and Molecular Biology, Harvard University (L.C.), Cambridge, Massachusetts 02138

I. INTRODUCTION

The in vitro differentiation of murine erythroleukemia (MEL) cells provides a useful system in which control of the differentiation program can be studied at the molecular level. MEL cells may be grown indefinitely in suspension but become committed to terminal erythroid differentiation when treated with dimethylsulfoxide (DMSO) or a variety of other agents [Housman et al, 1978]. Once a cell is committed to terminal differentiation the inducer is no longer required for execution of this program [Gusella et al, 1976; Levenson and Housman, 1979; Housman et al, 1978]. The commitment step thus represents

Levels of Genetic Control in Development, pages 171–183

a crucial transition for each cell in initiating the differentiation process. To measure the proportion of cells in a culture that has undergone the transition to the committed state, we have employed a method in which cells are exposed to inducer in liquid culture and then cloned in a semi-solid medium (plasma clot) in the absence of the inducer. Two basic colony phenotypes are observed. Cells that have not yet committed to inducer-independent erythroid differentiation (uncommitted cells) give rise to colonies in which all cells stain negatively with the heme-specific benzidine stain. Progeny of an uncommitted cell continue to proliferate, and colonies derived from such cells can reach macroscopic size. However, cells committed to expression of the terminal differentiation program give rise to colonies with a very different phenotype. All cells in such a colony have high levels of hemoglobin and stain intensely with benzidine. The proliferative capacity of a committed cell is limited and gives rise to colonies with a maximum of 32 cells. The cell nuclei have highly condensed chromatin, in contrast to the diffuse chromatin of the undifferentiated MEL cell [Housman et al, 1978]. We observed a very high degree of correlation between these phenotypes, suggesting that a single reprogramming event is responsible for the initiation of hemoglobin synthesis, the programmed loss of proliferative capacity, and the condensation of the chromatin.

Our interest has focused on events occurring during the latent period, prior to the point at which a significant number of MEL cells become committed to terminal erythroid differentiation. A number of studies have suggested that changes in the cell membrane are the earliest detectable events induced by DMSO. It has been shown that changes in rates of transport of various molecules are altered during the earliest period of exposure to DMSO [Mager and Bernstein, 1978]. The mean cell volume is also reduced during this period [Loritz et al, 1977].

Recent studies have suggested that changes in cation transport may play an important role in the early events in the differentiation program. Bernstein and co-workers [1976] have shown that exposure of MEL cells to ouabain causes differentiation of the cells. This result suggests that the sodium, potassium ATPase may play a role in the control of the differentiation process. Exposure of MEL cells to high levels of K^+ induced erythroid differentiation as well [Mager et al, 1979]. This result can also be interpreted to indicate the importance of potassium ion fluxes in the control of MEL cell differentiation. These results led us to explore further the role of ion fluxes in the control of the differentiation program of MEL cells. We have therefore chosen to analyze the differentiation process by varying the concentrations or rates of uptake of specific ions. We have found that, in particular, variation in the calcium concentration of the medium or the rate at which calcium is taken up by the cells has a dramatic effect on the rate at which cells become committed to terminal erythroid differentiation [Levenson et al, 1980].

II. RELATIONSHIP OF ION TRANSPORT TO MEL CELL DIFFERENTIATION

A. Effect of EGTA on Commitment

In an initial series of experiments, we analyzed the effect of the calcium chelating agent EGTA on MEL cell differentiation. Cells were treated with inducer (1.5% DMSO) under conditions where free calcium levels were reduced to submicromolar levels (2.7 mM EGTA). Exposure of cells to EGTA for 36 hours resulted in only a slight inhibition of cell growth (<10%), whereas commitment to erythroid differentiation was dramatically affected. It is important to note that the efficiency of colony formation was close to 100% in all experiments. This ensures that during the experimental treatment period the cells do not suffer severe cytotoxic or irreversible damage due to drug treatment or an adverse ionic environment. As illustrated in Figure 1, addition of EGTA to differentiating cells at various times during the latent period prevented the cells in the culture from undergoing the commitment process. Commitment was rapidly and effectively blocked if EGTA was added to DMSO-treated cells at 24 hours, a point after the end of the latent period. The inhibition of MEL cell differentiation caused by EGTA could be reversed if free calcium (1–2 mM; C_F) was added to EGTA-treated cultures. This suggests that the inhibitory effect caused by EGTA is in fact due to chelation of calcium ions rather than some other effect of the EGTA molecules.

B. Relationship of Calcium Flux to Commitment

The results of these studies led us to the view that the intracellular level of calcium ions might play an important role in the differentiation of MEL cells, and in particular, in the commitment process itself. Further support for this hypothesis came about somewhat serendipitously in our work using the drug amiloride [Levenson et al, 1980]. Our initial intent in employing this agent was to examine the role of sodium transport in MEL cell differentiation, since Bernstein and co-workers [1976] had shown that ouabain, an inhibitor of the plasma membrane (Na^+, K^+)-ATPase, could serve as an inducer of erythroid differentiation in this system. We wished to examine the potential role of sodium transport in greater detail, and planned to do so by analyzing the effect of an inhibitor of sodium transport on the differentiation process. Amiloride was chosen because it has been shown to be a reversible inhibitor of sodium transport in a variety of other systems [Baer et al, 1967; Bentley, 1968; Aceves and Cereijido, 1978; Koch and Leffert, 1979].

Our initial experiments with this drug indicated that at a dose of amiloride that had virtually no effect on cell growth (10μg/ml) the commitment of cells

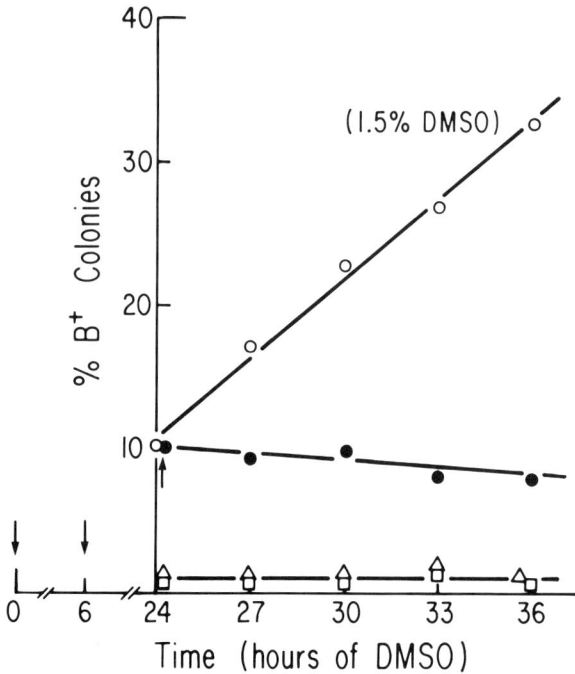

Fig. 1.   Effect of EGTA on MEL cell commitment. MEL cells were grown in liquid culture at a density of 1–4 × 10⁵ cells/ml in the presence of 1.5% DMSO (O—O). At various times, cells were subcultured, and growth continued in the presence of 1.5% DMSO + 2.7 mM EGTA. EGTA was added to subcultures at 0 hours (□—□), six hours (Δ—Δ) or 24 hours (●—●) after DMSO addition (arrows). At the times indicated, an aliquot of cells was removed from each culture and plated in plasma culture in the absence of DMSO or EGTA. Plasma cultures were harvested 96 hours later, and the proportion of heme-containing (benzidine-positive) colonies was determined as previously described. (Gusella et al, 1976; Levenson and Housman, 1979)

to erythroid differentiation could be completely blocked. As shown in Figure 2, this inhibition could be achieved if cells were treated with drug prior to, at the time of, or up to six hours after exposure to DMSO. Addition of amiloride after six hours of DMSO treatment had virtually no inhibitory effect on the commitment process. To our surprise, however, when we analyzed the action of the drug on ion transport, we could detect no significant effect of amiloride on cytoplasmic sodium levels.

While these experiments were in progress, A. Tsiftsoglou in our laboratory had found that the membrane active drug procaine also was an effective inhibitor of MEL cell differentiation. He found that the inhibitory effect of this agent could be reversed if free calcium was added to the medium of procaine-treated

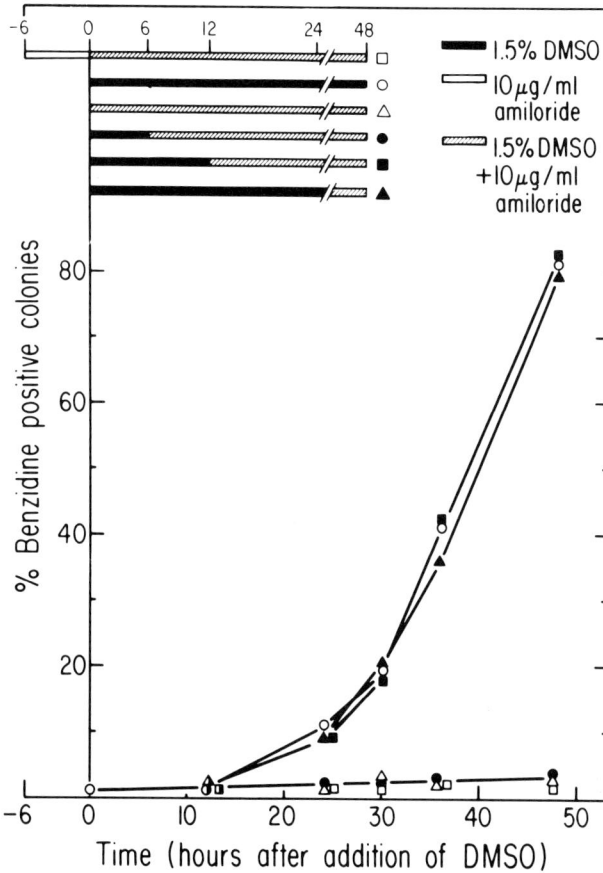

Fig. 2.   Effect of amiloride on MEL cell commitment. MEL cells were grown in liquid culture in the presence of 1.5% DMSO. At various times, cells were subcultured and were treated as follows: one culture was grown in the presence of 1.5% DMSO for the entire experiment (O—O), and a second culture was treated continuously with 1.5% DMSO + 10 μg/ml of amiloride (Δ—Δ). Amiloride (10 μg/ml) was added to subcultures at either six hours (●—●), 12 hours (■—■), or 24 hours (Δ—Δ) after treatment with DMSO. One culture (□—□) was treated with 10 μg/ml amiloride for six hours, then grown in the presence of 1.5% DMSO + 10 μg/ml amiloride for the remainder of the experiment. At the times indicated, an aliquot of cells was removed from each culture and plated in plasma culture in the absence of DMSO or amiloride. Culture conditions, harvesting, and scoring of colony type were as in Figure 1.

cells. Since amiloride shared some similarities with procaine, we reasoned that addition of calcium to amiloride-treated cells might also reverse the block to differentiation. Although addition of calcium did not reverse the amiloride effect, we found that amiloride inhibition of MEL cell commitment was reversed by treatment of cells with the calcium ionophore A23187 (1 μg/ml), as shown in

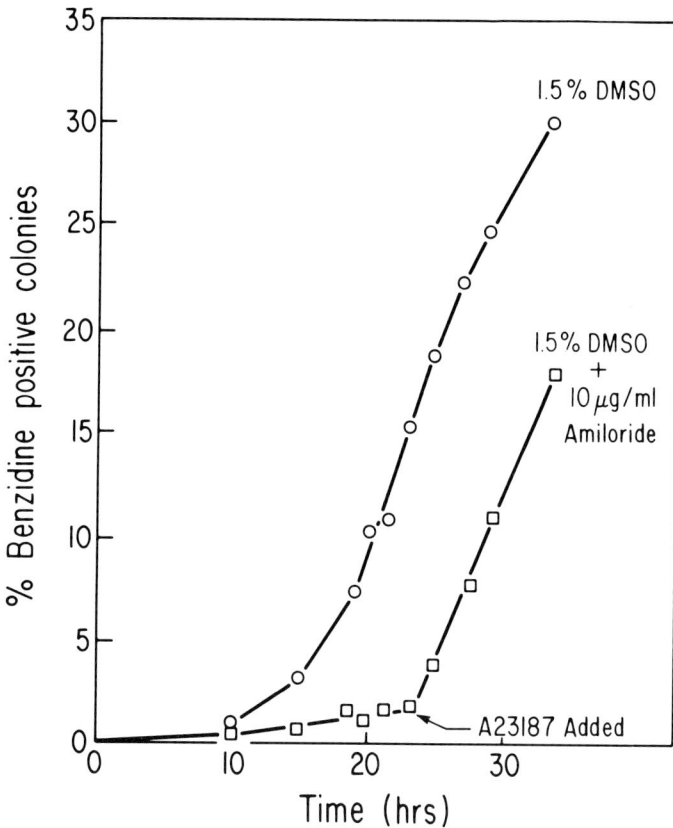

Fig 3. Reversal of amiloride inhibition. MEL cells were grown in liquid culture in the presence of 1.5% DMSO alone (O—O) or 1.5% DMSO + 10 μg/ml of amiloride (□—□). At 24 hours, A23187 (1 μg/ml) was added to the amiloride-treated culture, and growth continued in the presence of DMSO, amiloride, and ionophore. At the times indicated, aliquots of cells were plated in the absence of DMSO, amiloride, and ionophore. Plasma culture conditions and determination of benzidine-positive colonies were as described in Fig. 1.

Figure 3. It should be pointed out that treatment of MEL cells with A23187 alone was not sufficient to induce differentiation.

Our experiments with EGTA and amiloride suggested, therefore, that calcium ions played an important role in the commitment process. Direct measurements of calcium levels in MEL cells under various conditions have confirmed this hypothesis [Levenson et al, 1980]. Cells treated with 1.5% DMSO exhibit a significant increase in both the rate of calcium transport and the intracellular level of calcium ions compared with control cells. Cells treated with DMSO and amiloride, however, exhibit calcium uptake rates and calcium levels similar to those of control cells.

III. RELATIONSHIP BETWEEN TRANSMEMBRANE POTENTIAL AND MEL CELL DIFFERENTIATION

A. Uptake of Lipophilic Cations

The central role played by calcium in regulating a variety of cellular functions has led us to investigate the relationship between the commitment of MEL cells to terminal differentiation and several cellular parameters that are known to be related to calcium flux. One such parameter is transmembrane potential, changes in which have been postulated to mediate a variety of physiologic and metabolic responses in cells [Kiefer et al, 1980; Korchak and Weissmann, 1978; Lantz et al, 1980; Gallin and Gallin, 1977]. One approach we have initiated to address this issue has been to measure changes in membrane potential as a function of time of inducer treatment employing the membrane potential sensitive lipophilic cation triphenyl methyl phosphonium+ (TPMP+). The partitioning of TPMP+ between cells and the medium is a function of membrane potential [Schuldiner and Kaback, 1975]. The uptake of this indicator can be easily quantitated and provides a rapid estimation of the average membrane potential of cells: the more indicator taken up by the cells, the greater the hyperpolarization of cells due to an increase in the net negative charge inside the cell with respect to the exterior. The results of our initial experiments, shown in Figure 4, suggest that ^{3}H-TPMP+ incorporation is related to the differentiation state of cells during induction. The rates of incorporation and intracellular levels of TPMP+ in a control culture, a culture treated with 1.5% DMSO for 15 minutes, and a culture treated with 1.5% DMSO for 14 hours were found to be virtually identical. The intracellular level of TPMP+ in a culture treated with 1.5% DMSO for 24 hours, however, was reduced by >40%. This result suggests that treatment of cells with DMSO causes a significant membrane depolarization, and that this change may occur at or near the time of commitment. It has not yet been resolved, however, whether the depolarization we have observed occurs at the plasma membrane or in the membrane of organelles such as mitochondria. Further experiments will be needed to resolve this issue.

B. Flow Cytometry

While ^{3}H-TPMP provides a rapid estimation of the average membrane potential of the cells in culture, it does not allow detection of heterogeneity within the cell population. This is of significance since we have previously shown that individual MEL cells enter the differentiation program in an asynchronous fashion. Fluorescent dyes, coupled with the technique of flow cytometry, can be used to estimate the membrane potential of individual cells [Shapiro et al, 1979]. Our initial experiments, carried out with the fluorescent dye 3,3'-dihexyloxa-

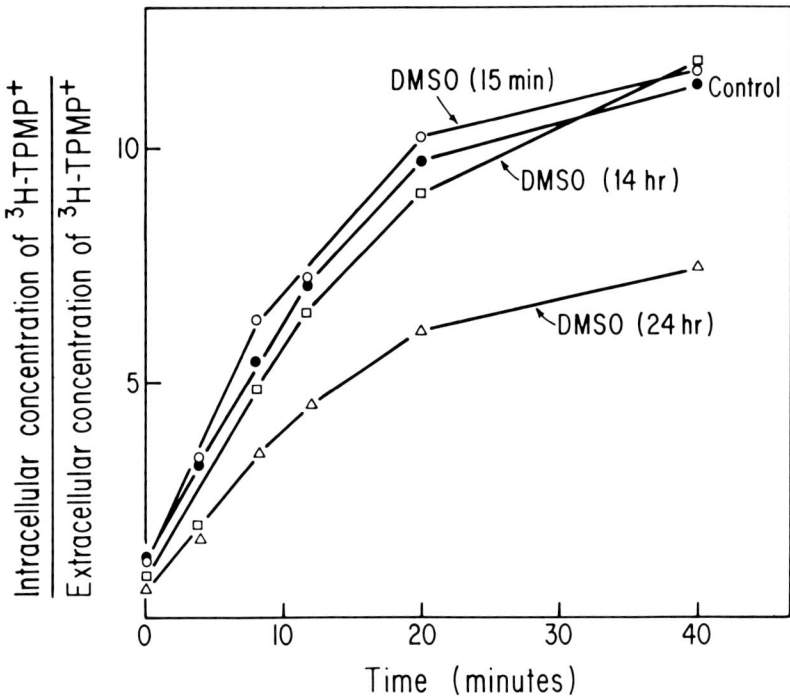

Fig. 4 Effect of DMSO on membrane potential. MEL cells were grown in liquid culture under the following conditions: ●—● untreated cells; □—□ 14 hours of treatment with 1.5% DMSO; Δ—Δ 24 hours of treatment with 1.5% DMSO. A fourth culture (○—○) was treated with 1.5% DMSO for 15 minutes. Cells were concentrated to a density of 3.5 × 10⁶ cells/ml in 10 ml of culture media. 20 λ of ³H-TPMP⁺ (5 × 10⁵ cpm/λ) was then added to 7 ml of cells. At 0 time and at the intervals shown, 1 ml aliquots of cells were removed from each culture and centrifuged, and the radioactivity in the cell pellet and supernatants was determined. The amount of cell water was determined by adding a trace amount of ³H₂O and [¹⁴C]-sucrose to the remaining 3 ml of cells. Isotopic equilibrium was reached after 15 minutes. 1 ml aliquots of cells were then removed from each culture and centrifuged, and the radioactivity in the cell pellet and supernatants was determined. Cell concentrations were determined using an automatic cell counter. The intracellular level of water was determined by subtracting the amount of water made available to sucrose from the total water in the cell pellet as described in Levenson et al [1980].

carbocyanine iodide [DiOC₆-(Eastman)], indicated the utility of this approach as a method for detecting membrane potential changes in individual cells and as a possible method for obtaining populations of cells in various stages of the differentiation program.

   The fluorescence pattern of MEL cells treated with DiOC₆ is shown in Figure 5. Treatment of MEL cells for 24 hours with 1.5% DMSO causes a significant reduction in dye uptake compared to cells untreated with DMSO. This reduction

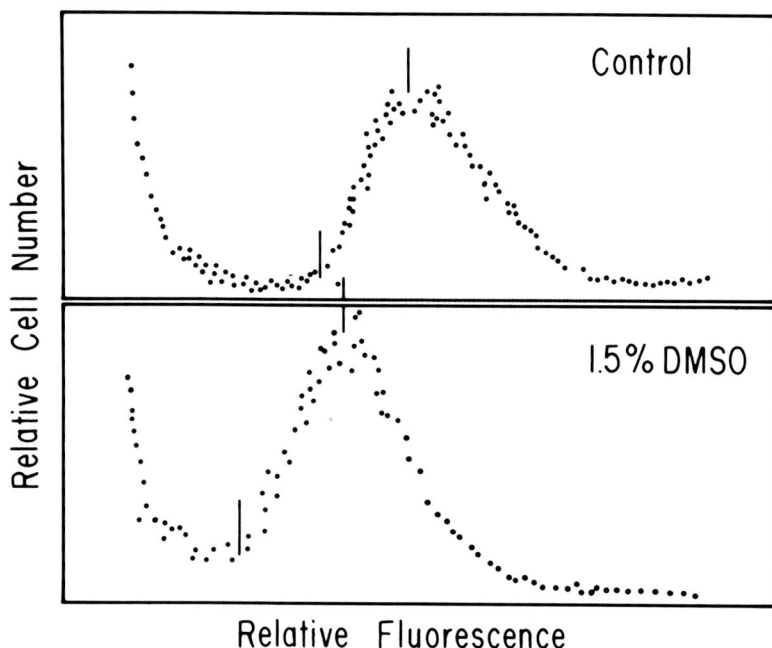

Fig. 5. Relative changes in fluorescence of $DiOC_6$ in MEL cells induced to differentiate with DMSO. MEL cells were grown in liquid culture at a density of $1–4 \times 10^5$ cells/ml for 24 hours in the presence of 1.5% DMSO. Cells were adjusted to a density of 5×10^5 cells/ml in media lacking fetal calf serum but containing 1.5% DMSO. 10 μl of a 5 μM solution of $DiOC_6$ in ethanol was added per ml of cells as described by Shapiro et al [1979]. Cells and dye were allowed to equilibrate for 30 minutes at room temperature. After incubation, samples were analyzed in a FACS II-cell sorter (Becton-Dickinson) using excitation at 488 nm from an argon laser source. The distributions of fluorescence from individual cells were accumulated in a multichannel pulse height analyzer.

in dye uptake is indicative of a decrease in membrane potential in response to DMSO, in agreement with our previous data obtained with $TPMP^+$.

These results suggest that the change in membrane potential caused by DMSO treatment may play an important role in the commitment process. The use of membrane potential sensitive fluorescent dyes and flow cytometry, coupled with cell sorting techniques, may in the future provide both an analytic and a preparative method for obtaining pure populations of cells in different stages of the differentiation program.

IV.   THE CYTOSKELETON AND MEL CELL DIFFERENTIATION

Our work on the relationship between ion flux and the commitment process has led us to consider the role of the cytoskeleton in MEL cell differentiation.

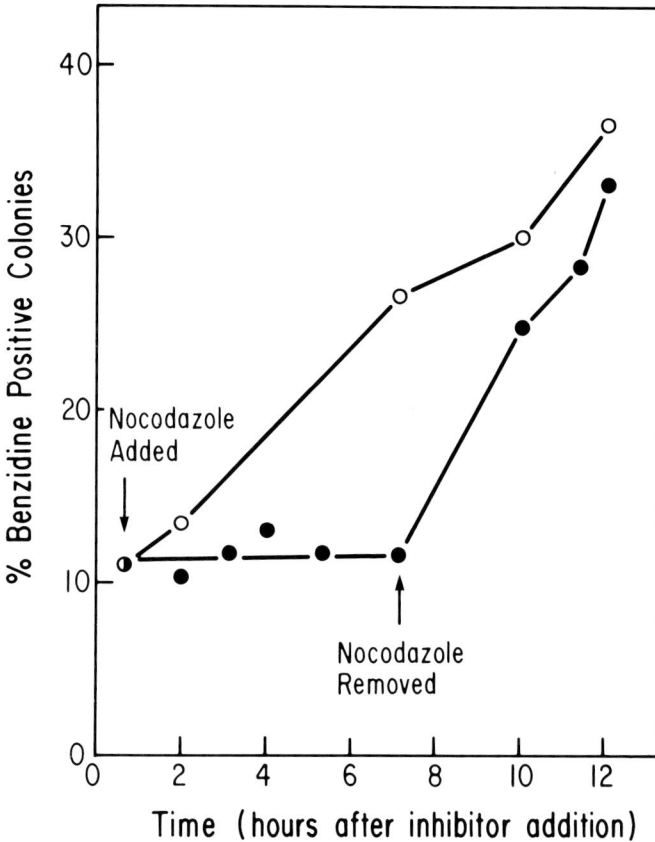

Fig. 6.   Effect of nocodazole on MEL cell differentiation. MEL cells were grown in liquid culture in the presence of 1.5% DMSO for 24 hours. At this time, cells were subcultured and growth continued either in 1.5% DMSO alone (O—O) or 1.5% DMSO + 0.1 μg/ml nocodazole (●—●). After six hours of drug treatment, nocodazole was removed, and growth was continued in 1.5% DMSO alone for the remainder of the experiment. At the times indicated, an aliquot of cells was removed from each culture and plated in plasma culture in the absence of DMSO or nocodazole. Plasma culture, harvesting, and scoring of colony type were as described in Fig. 1.

Cytoplasmic microtubules constitute a major portion of the cytoskeleton. The distribution and organization of the microtubules apparently is sensitive to the level of cytoplasmic calcium [Poste and Nicholson, 1976]. Variation in cytoplasmic calcium levels thus can influence the distribution of microtubules in the cytoplasm, leading to changes in cell morphology, movement, and other functions.

The fact that treatment of MEL cells with DMSO leads to a significant influx of calcium ions suggests that alterations in microtubular organization may occur in response to this influx. We therefore wished to examine if perturbations of microtubule organization have an effect on the commitment of cells to terminal differentiation.

Our approach has been to analyze the effect of agents that depolymerize microtubules on the ability of cells to undergo the commitment process. In these initial studies, we have used the colchicine analog nocodazole, a drug that specifically depolymerizes microtubules [Solomon, personal communication]. Our preliminary studies suggest that nocodazole has a dramatic effect on the rate at which cells become committed to terminal erythroid differentiation.

The results of an experiment employing nocodazole are shown in Figure 6. Cells were treated with 1.5% DMSO for 24 hours, so that approximately 11% of the cells in culture were committed. At this point nocodazole ($C_F = 0.1$ γ/ ml) was added to cells in the continued presence of DMSO, and cells were grown for an additional six hours in the presence of both drugs. Cells were then washed free of nocodazole and grown in 1.5% DMSO for the remainder of the experiment. As can be seen in Figure 6, nocodazole rapidly prevented any further cells from undergoing the commitment process. Upon drug removal, cells reinitiated the differentiation program with little or no lag.

These results are provocative in that they imply that the organization of the microtubule system is in some way related to the execution of the differentiation program. Owing to the fact that MEL cells are spherical and do not attach well to inert supports, it has been difficult to visualize directly the cytoskeleton and any changes that might occur in the cytoskeleton organization during differentiation. Direct biochemical examination of microtubule preparations from nocodazole-treated and control cells have been inconclusive in demonstrating a direct effect of the drug on cytoplasmic microtubules. Nocodazole is also known to cause cells to accumulate in mitosis [Zieve et al, 1980]. Our results do not permit us as yet to distinguish whether nocodazole blocks MEL cell differentiation via a reorganization of cytoplasmic microtubules or whether the drug prevents commitment by virtue of its effect on a cell cycle-related process. Nonetheless, the fact that nocodazole reversibly inhibits the commitment of MEL cells to differentiate may provide an important clue as to how the commitment process is regulated at the molecular level.

V. SUMMARY

Our recent work has focused on the relationship between ion flux and the commitment of MEL cells to the program of terminal erythroid differentiation. We have found that alterations in the calcium concentration of the medium has a profound effect on the rate at which cells become committed. DMSO has been found to cause a significant increase in the rate at which calcium ions are taken up by cells, as well as causing an increase in the level of cytoplasmic calcium. The uptake of calcium by cells in response to DMSO appears to be a critical event in the commitment process, since amiloride, a drug that inhibits commitment, also inhibits the uptake of calcium. The inhibition of commitment caused by amiloride can be overcome by treatment of cells with the calcium ionophore A23187.

We have now begun to investigate whether cellular processes known to be affected by calcium ions are related to the commitment process. In particular we have analyzed the relationship between changes in both membrane potential and the organization of cytoplasmic microtubules to the commitment of cells to terminal differentiation. We have employed the membrane potential sensitive lipophilic indicators ^{3}H-TPMP$^+$ and the fluorescent cation dye DiOC$_6$ to assess the relationship between commitment and changes in membrane potential levels. Our results indicate that in response to DMSO, cells undergo a reduction in membrane potential levels that may be temporally related to the commitment event. Alterations in membrane components thus may be of importance in the differentiation process. Finally, a possible relationship between the organization of the cytoskeleton and the commitment process has been suggested by our work with the drug nocodazole.

Our results suggest that the program of MEL cell differentiation leading up to and including the commitment process itself comprises a complex series of events that include alterations in ion transport and changes in membrane components. How these parameters are involved in, and possibly regulate, the dramatic changes we observe in the developmental potential of the cell is not yet known.

ACKNOWLEDGMENTS

We thank Dr. E.J. Cragoe, Jr., Merck, Sharp and Dohme, West Point, PA. for his generous gift of amiloride. We also thank Dr. Efraim Racker for his advice on calcium measurement techniques. We are extremely grateful to Dr. Frank Solomon for his continuing help and suggestions throughout the course of this work. We are also grateful to Dr. Sam Latt for the use of his FACS cell sorter and instruction in its use. This work was supported by grant CA17575 from the NIH (D.H.) and grant 791040 from the American Heart Foundation (L.C.).

REFERENCES

Aceves J, Cereijido M (1978). J Physiol 229:709–718.

Baer JE, Jones CB, Spitzer AS, Russo HF (1967). J Pharmacol Exp Ther 157:472–485.

Bentley PF (1968). J Physiol 195:317–330.

Bernstein A, Hunt V, Crichley V, Mak TW (1976). Cell 9:375–381.

Gallin EK, Gallin JL (1977). J Cell Biol 75:277–289.

Gusella JF, Geller R, Clarke B, Weeks V, Housman D (1976). Cell 9:221–229.

Housman D, Gusella J, Geller R, Levenson R, Weil S (1978). In "Differentiation of Normal and Neoplastic Hematopoietic Cells." (Clarkson, Marks, and Till, eds.), pp 193–207. Cold Spring Harbor Laboratory, New York.

Kiefer H, Blume AJ, Kaback HR (1980). Proc Natl Acad Sci USA 77:2200–2204.

Koch KS, Leffert HL (1979). Cell 18:153–163.

Korchak HM, Weissmann G (1978). Proc Natl Acad Sci USA 75:3818–3822.

Lantz RC, Elsas LJ, De Haan RL (1980). Proc Natl Acad Sci USA 77:3062–3066.

Levenson R, Housman D (1979). J Cell Biol 82:715–725.

Levenson R, Housman D, Cantley L (1980). Proc Natl Acad Sci USA 77:5948–5952.

Loritz F, Bernstein A, Miller RG (1977). J Cell Physiol 90:423–435.

Mager D, Bernstein A (1978). J Cell Physiol 94:275–285.

Mager DL, MacDonald ME, Bernstein, A (1979). Dev Biol 70:268–273.

Poste G, Nicolson GL (1976). Biochim Biophys Acta 426:148–155.

Schuldiner S, Kaback HR (1975). Biochemistry 14:5451–5461.

Shapiro HM, Natale PJ, Kamentsky LA (1979). Proc Natl Acad Sci USA 76:5728–5730.

Zieve GW, Turnbull D, Mullins JM, McIntosh TR (1980). Exp Cell Res 126:397–405.

Co-Evolution and Control of Globin Genes

Oliver Smithies, Ann E. Blechl, Sixiang Shen, Jerry L. Slightom, and Elio F. Vanin

Laboratory of Genetics, University of Wisconsin, Madison, Wisconsin 53706

I. INTRODUCTION

Recent successes in cloning and sequencing long stretches of DNA from the genomes of mammals have resulted in a great deal of detailed information at the nucleotide level, particularly about the globin genes. From this information one can draw a number of interesting new inferences concerning the evolution and co-evolution of the globin genes. One can also begin to see indications of unexpected genetic elements within the globin system which may be important in controlling the expression of the globin structural genes at different developmental stages. In discussing these various new findings and the ideas that stem from them, I shall of course use data obtained in our laboratory, but I shall also need to draw on work from many other groups.

A closely related version of this paper is published in the Second Conference on Hemoglobin Switching; Stamatoyannopoulos G, Nienhuis AW (eds): *Organization and Expression of Globin Genes,* New York: Alan R. Liss, Inc., 1981.

Levels of Genetic Control in Development, pages 185–200

Alpha

Beta

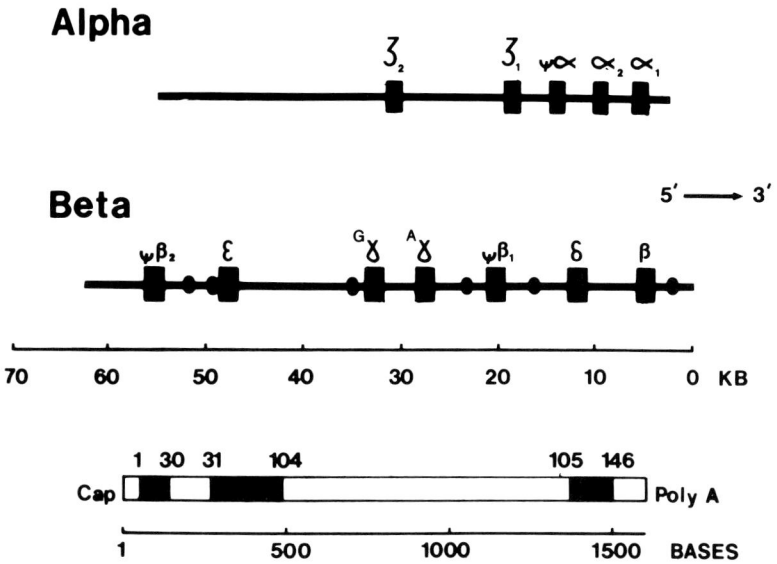

Fig. 1. The human α and β globin gene clusters (Modified from Efstradiatis et al [1980]). The genes of the human alpha and β globin gene loci are shown as black bars. The solid circles indicate the distribution of some repeated DNA sequences in the β gene cluster [Fritsch et al, 1980]. The arrangement of coding and intervening sequences in the $^G\gamma$ fetal globin gene is shown on an enlarged scale; the numbers refer to amino acid residues, and the solid bars represent the amino acid coding regions of the gene.

II.   ANATOMY OF THE GLOBIN GENE CLUSTERS AT THE DNA LEVEL

Figure 1 [modified from Efstradiatis et al, 1980] summarizes much of what is known about the general disposition of the human α and β globin gene clusters on chromosomes 16 and 11 respectively.

It is important to recognize that 1) each cluster contains several genes; 2) they are all relatively close together on their respective chromosomes [the scale is such that the line representing the DNA of the β region is about 60,000 base pairs (60 kbp); the α cluster is about 40 kbp]; 3) the genes are all oriented in the same direction relative to transcription with 5' to the left and 3' to the right, so that transcription is left to right; 4) all the human globin polypeptides (α, β, γ, δ, ε, and ζ) are accounted for and the α, γ, and ζ polypeptides are coded for by duplicate genes. 5) in addition to the genes coding for known polypeptides, there are some extra genes in these clusters, which I shall refer to as "pseudo-genes"; one is known in the α cluster and two are known in the β cluster. 6) between many, but not all, of the genes there are some repeated sequences of DNA (vide infra). 7) finally, the genes are arranged on the chromosome in the order in which they are used developmentally, for example in the β globin gene cluster the embryonic globin gene, ε, is 5' to the fetal globin genes, $^G\gamma$ and $^A\gamma$, and these are 5' to the adult globin genes δ and β; similarly, the embryonic ζ genes are 5' to the adult α genes.

III. EVOLUTION OF GLOBIN GENES

Amino acid sequence data available for many years [Dayhoff, 1972] have established that the globins evolved from a common ancestor by a process of gene duplication and subsequent evolution. Consequently it is not surprising that the recent data at the DNA level show that all of the genes that yield a globin product (ie, excluding the pseudogenes) have the same general anatomy, which is illustrated in the case of the $^G\gamma$ gene in the lower part of Figure 1. For example, all the productive genes have two intervening sequences (or introns): the first between the codons for amino acid residues 30 and 31 and the second between residues 104 and 105. Both the α- and β-type globin genes have their intervening sequences in homologous positions, which leads to the likely inference that the common ancestor of the α and β genes had already acquired intervening sequences, as was originally pointed out by Leder et al [1978].

The genes for the α- and β-type globins are located on separate chromosomes in most species, and it is not surprising that although they once had a common ancestor they have now diverged considerably. An example of this divergence at the DNA level is seen in the size of the second intervening sequence. In all sequenced β-type globin genes the second intervening sequence is many times larger than the first, whereas in the α-type globin genes the two are of similar size.

The complete DNA sequences of all the human β-type globin genes are now available, and the laboratories involved in sequencing these genes (including our own) have collaborated in a joint paper comparing the sequences [Efstradiatis et al, 1980]. A rather accurate and detailed evolutionary tree was constructed from these sequences by comparing nucleotides that are important in coding, but excluding silent nucleotides (such as nucleotides in the introns and many of the third bases in coding triplets). This tree is reproduced in Figure 2, with a few changes mainly to indicate the presence of the pseudogenes. As yet there is insufficient evidence to allow me to place the pseudogenes accurately, but putting them into the tree in a preliminary way is nevertheless a helpful reminder that their presence must be taken into account.

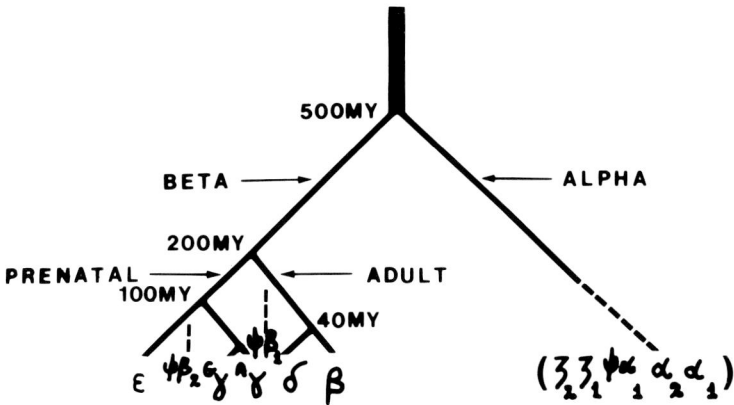

Fig. 2. An evolutionary tree for the human globin genes (modified from Efstradiatis et al [1980]). The tree was constructed by direct comparisons of the nucleotide sequences of the coding regions of the respective globin polypeptides, excluding silent bases and intervening sequences. The dotted lines indicate parts of the tree that have not yet been ascertained. MY = millions of years.

IV. GENERAL ASPECTS OF GENE DUPLICATIONS

The evolutionary tree leads me to the major aspect of the evolutionary part of my presentation. I have been interested for many years in what goes on at the branch points of trees such as this; ie, what is the nature of the event that leads from a single copy of a gene to two copies. (My interest in this topic dates back 20 years to work I did on a partial gene duplication in the human haptoglobin genes [reviewed in Smithies, 1964]). If we take the simplest interpretation of the globin evolutionary tree, it shows that for the β type globins the first surviving duplication event occurred about 200 million years ago.

Let us consider this event in general terms. Figure 3A diagrams the series of events classically thought to lead to a de novo duplication. You will see that the extent of a duplication is likely to be considerably larger than the region actually coding for the specific gene product being studied. This means that duplicated genes will usually be surrounded by regions of DNA that are also duplicated. Notice that the true length of the duplication is given by the distance between homologous points (eg, the distance from the E to E) rather than by the intergenic distance (I to F). Figure 3A also shows that it may be possible to deduce the type of event that led to a given duplication by examining the DNA sequences from three particular regions. In the duplication of Figure 3A these three regions are the CD region (which existed prior to the duplication), the JD region (which is the "novel joint"), and the other preexisting region, JK.

A.

A B C D E F G̅H̅I J K L M N O

A̲B̲C̲D E F G̅H̅I J K L M N O

▼

A B C D E F G̅H̅I J D E F G̅H̅I J K L M N O

B.

A B C̲D̲ E F G̅H̅I J̲D̲ E F G̅H̅I J̲K̲L̲ M N O

▼

A B C̲D̲ e F g̅H̅l ̲j ̲d E f G̅h̅l J̲K̲L m N o

C.

A B C D e F g̅H̅l j d E f G̅h̅l J K L m N o
                    X
    A B C D e F g̅H̅l j d E f G̅h̅l J K L m N o

▼

A B C D e F g̅H̅l j d e F g̅H̅l j d E f G̅h̅l J K L m N o
                    or
A B C D E f̅G̅h̅l J K L m N o

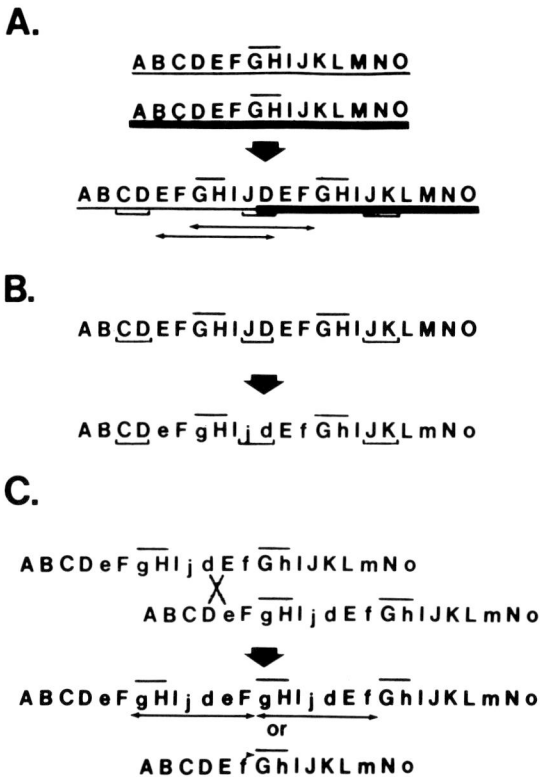

Fig. 3.   De novo gene duplications and their subsequent evolution. A. A schematic representation of the de novo formation of a duplication of a gene "GH" by the breakage and reunion of identical DNA molecules at non-homologous positions, CD and JK. The brackets indicate the regions that require study in order to determine the nature of the breakage and reunion event. The horizontal arrows illustrate that the length of the duplicate regions can be determined from the distance between many pairs of homologous positions (eg, from the E to E distance). B. The divergence of duplicated genes by mutations subsequent to the initial duplication is illustrated. A change from large to small letters represents observable sequence differences that do not obscure the relatedness of two regions. Note that comparisons of the CD, jd, and JK sequences can still show the site of breakage and reunion. C. Unequal but homologous crossing over can lead to gene triplication or to gene loss. The horizontal arrows show that the spacing of the triplicate genes is the same as between the duplicates.

Once a duplication exists in the genome, mutations will occur that will cause the initial duplicates to diverge. Consequently it may also be possible to deduce what has happened to a duplication since its initial formation by comparing the subsequent divergence of the duplicated regions of DNA, which were identical when the duplication first arose (Fig. 3B). In addition, secondary unequal but homologous crossover events may be expected to lead to an increase (or decrease) in gene number without the need to invoke anything more unusual than recombination between homologous (duplicated) DNA sequences (Fig. 3C). Products of this secondary increase in gene number have some quite characteristic features, such as the fact that the now triplicated genes are uniformly spaced and that each will have very similar surroundings. This last feature should be useful in distinguishing genes that have arisen by secondary events from those that have arisen by a de novo duplication. Thus, in Figure 3C, the triplicated genes have between them the two very similar sequences IjdeF and IjdEf, but outside the triplicated region two quite different sequences occur: ABCDeF and IJKLmNo.

V. DUPLICATION OF HUMAN FETAL GLOBIN GENES

One of the aims of our work is to describe the evolution of the β-globin gene cluster in terms of de novo duplications and subsequent secondary events. But you will realize from the complexity of the region that its evolution is not likely to be simple. Not only do we have to account for the nature and distribution of seven globin genes (I include the pseudogenes in this count), but in addition we have to explain the positions in the region of at least six copies of a repeated DNA sequence. We have tried to assemble a detailed evolutionary path compatible with all the details of the known map but have not yet been successful because there are insufficient data concerning the similarities and differences within the intergenic regions. However, we did much better when we confined our attention to the fetal globin gene region, which we have almost completely sequenced.

The length of the duplication that created the two human fetal globin genes is approximately 5 kbp; we know this because the distance between homologous points in the two genes is 5 kbp (see Fig. 3A). A complete comparison of the duplicate regions would therefore need more than 10,000 bp of DNA sequence data. Figure 4 presents a comparison of the data we have so far accumulated— over 7,000 bp of sequence. The vertical bars show base pair differences per 100 base pairs plotted as a function of their position along the duplication.

The results of the comparison are to a large extent as predicted. The general location of the 5' end of the duplication is easily identified. To the left (5') of the point of duplication, the sequence of the $^G\gamma$ gene (BC in Fig. 3) deviates completely (> 70% difference) from the sequence in the comparable position 5' to the $^A\gamma$ (Ij in Fig. 3). Yet to the right or 3' side of the point of duplication the sequences of the $^G\gamma$ and $^A\gamma$ genes are very homologous, thus the $^G\gamma$ sequence (de in Fig. 3) is identical in sequence to the $^A\gamma$ sequence (dE in Fig. 3) in more than 65% of its nucleotides.

We are, of course, very interested in finding whether there is anything unusual about the DNA in this region which might explain why the duplication end-

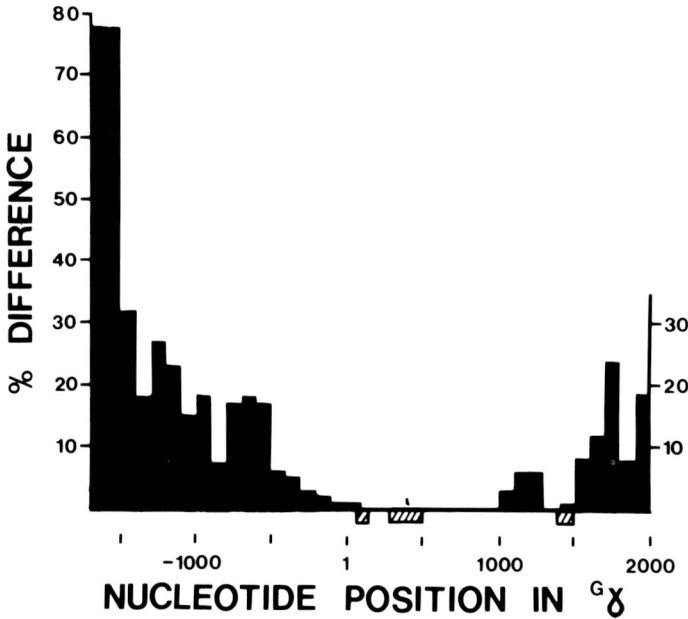

Fig. 4. A comparison of the duplicated $^G\gamma$ and $^A\gamma$ fetal genes and their surroundings. Vertical bars show the differences for every 100 base pairs (% difference) in the nucleotide sequences of the $^G\gamma$ and $^A\gamma$ duplicate genes plotted against positions along the duplication. Position 1 corresponds to the first nucleotide in $^G\gamma$ globin mRNA. The coding regions are indicated by hatched areas below the line. To calculate the % difference, single base substitutions, additions or deletions were counted as 1 difference; gaps greater than two base pairs were counted as 3 differences.

points are in these places. For example, we want to know whether the sequence CD has some unusual feature in common with the sequence JK that might account for the breakage and reunion occurring at these particular places. Until we have located and sequenced the right end of the duplication (JK) we cannot completely answer this question. Nevertheless we can say, even at this stage of our investigation when we have only sequenced CD and jd but not yet JK, that the region near the breakpoint is AT-rich (> 75% AT). Our current supposition is that the two fetal globin genes arose from a de novo duplication, that the two non-homologous breaks giving rise to the duplication occurred in AT-rich regions, and that the two AT-rich regions were sufficiently similar in sequence to permit the broken DNA molecules to re-join via their non-homologous ends. However we will not be able to prove the correctness of this supposition until we have the data for JK.

VI.   DNA EXCHANGES BETWEEN DUPLICATED GENES

A less expected feature of the comparison between the duplicated genes is a 1,200 bp stretch of DNA over which the $^G\gamma$ and $^A\gamma$ genes have only two base pair differences! This virtual identity contrasts markedly with the level of similarity (about 70–80%) between the other parts of the two genes. Clearly there must either be some strong selective factor that has prevented this region from diverging at the same rate as the remaining part of the duplication or some type of molecular mechanism exists that has kept this region identical in the duplicated genes on this chromosome (the region is not completely identical on the other chromosome from the same individual).

Once again, in order to look for a possible explanation at the molecular level for the existence of these non-divergent regions, we examined the DNA sequences at their boundaries. It was immediately apparent that the 3' boundary has some unusual features. In Figure 5 we have diagrammed the DNA sequences of the duplicated genes at and on either side of this boundary. The unusual feature is a string of TG dinucleotides (about two-thirds of the way into the larger intron) which form a stretch of simple sequence DNA. The length of this stretch of simple sequence DNA differs in the two duplicate genes in a way that looks like the result of an unequal but homologous crossing-over event. To the right of the simple sequence DNA the larger introns of the two genes differ in 12/269 bases; to the left, they are identical over a length of 571 bases. We have therefore proposed [Slightom et al, 1980] that this simple sequence DNA is a "hot spot" for initiating an intrachromosomal recombinational event, and that this event has led to an intergenic exchange of DNA. The result, in this instance, is that a large part of the $^A\gamma$ globin has been converted to be identical in sequence to that of the $^G\gamma$ globin gene. There is an indication that the conversion event shows some polarity (ie, it extends primarily to the 5' side of the "hot spot").

You will realize that the existence of gene conversion events can cause evolutionary trees, such as that presented in Figure 2, to be misleading. In many cases, this type of tree may not represent the history of the duplicational events so much as the history of the occurrences of gene conversions!

The type of event we think most likely to account for the intergenic exchange of DNA within a single chromosome in the human fetal globin genes is a sister chromatid exchange. (DNA sequence data from the other chromosome exclude its participation in this particular event.) Sister chromatid exchanges between duplicated genes can lead to daughters, which contain triplicate genes and a single gene, or they can lead to an intergenic exchange of DNA with no change in the number of genes in the two daughters. Figure 6 [from Slightom et al, 1980] illustrates the basis for this statement using intermediates in the recombinational event of the types currently proposed by workers in recombination [see Radding, 1978, for a review]. As you will appreciate, the figure accounts quite neatly for the data.

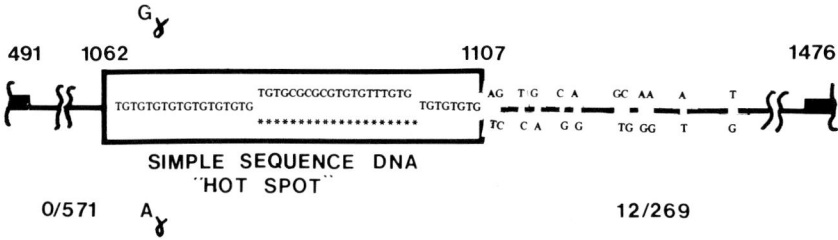

Fig. 5. A comparison of the DNA sequences of the larger introns of the duplicate $^G\gamma$ and $^A\gamma$ globin genes from one chromosone. The introns extend from nucleotide positions 491 to 1476 [Slightom et al, 1980]. The nucleotide sequences of the simple sequence DNA "hot spot" (see text) are shown in the box, with the asterisks indicating where the $^A\gamma$ gene is shorter. On the 5' side of the box the two introns are identical in sequence for 571 bases. On the 3' side they differ in the 12 (of 269) positions shown.

VII. PSEUDOGENES IN THE GLOBIN CLUSTERS

At this point I want to take up a different aspect of the globin gene clusters; namely, the nature of the pseudogenes to which I referred briefly in talking about Figure 1.

At present the DNA sequences of the human pseudogenes have not been published, although their orientation relative to the expressed genes is known, so I must turn to work on a mouse alpha globin-related pseudogene. About two years ago we described the isolation from the genome of outbred CDl Swiss mice of a clone, 30.5, which appeared to contain a gene related to the alpha globin gene family [Blattner et al, 1978; Smithies et al, 1978]. We have recently completed sequencing this gene [Vanin et al, 1980], and Nishioka et al [1980] have sequenced the comparable gene from Balb/c mice.

This alpha globin-related gene, 30.5, has several interesting features. First, the gene is clearly more related to the mouse alpha gene (84% identical over the coding region) than to the mouse beta genes (54% identity), and it is more closely related to the mouse than to the human or rabbit alpha globin genes. Second, it completely lacks both intervening sequences; they are as cleanly absent as if they had never been there (Fig. 7). Third, the gene could not code for a globin polypeptide, since it has frame shifts that produce missense amino acids and finally stop codons long before a normal globin could be produced (Fig. 8). Because of this third feature, we call it a pseudogene, $\psi\alpha$30.5. Its

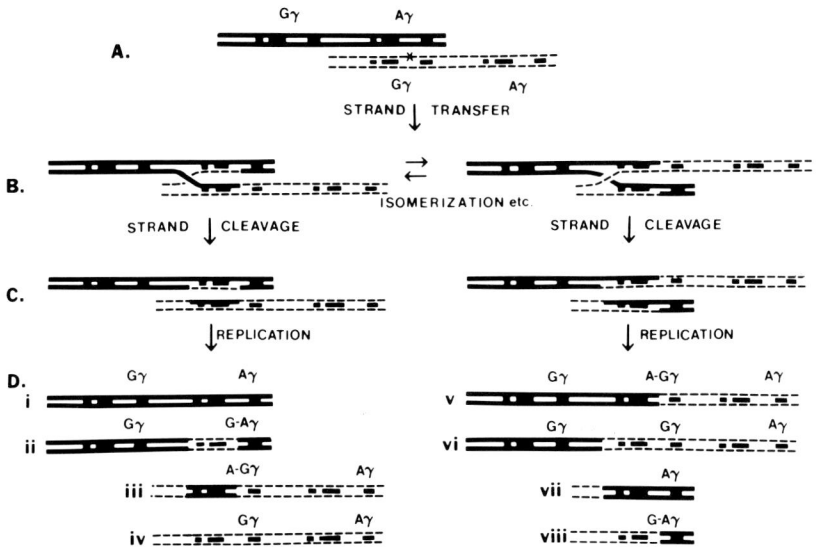

Fig. 6. Diagrammatic representation and possible outcomes of a recombinational event between a $^G\gamma$ globin gene aligned with an $^A\gamma$ globin gene (from Slightom et al [1980]). The participating DNA molecules are distinguished for illustrative purposes by light and heavy paired lines. Coding regions are shown by black bars between the paired lines. A. Initial alignment when the recombinational event was initiated at a simple sequence "hot spot" (X) in the larger intron. B. Strand transfer, isomerization, branch migration, and ligation yield the isomers. C. Strand cleavage leads to the intermediate heteroduplex products. D. Replication leads to the final products. Note that products ii and iii show "conversion."

second unusual feature also would have made it a pseudogene, because there is now a considerable body of data showing that if RNA is transcribed from globin (and other) genes without introns, then the RNA will not yield a stable cytoplasmic message [Hamer et al, 1979]. Thus the mouse alpha globin-related gene $\psi\alpha30.5$ is a nonproductive pseudogene in at least two senses.

How did this pseudogene evolve, and what is its function, if any? There are three general types of evolutionary history to be considered, none of which can be eliminated at present. The first evolutionary possibility is that the pseudogene is the modern descendant of an old precursor globin gene that lacked introns. This possibility is unlikely because the gene is so obviously related to the modern mouse adult alpha globin gene in the coding regions, but we cannot exclude it in view of our recent data showing that genes can exchange DNA along chromosomes. The second evolutionary possibility is that the introns were recently lost at the DNA level. We prefer this possibility, but we cannot prove that it is the best. A third possibility is that the pseudogene is derived from globin mRNA via the action of reverse transcriptase. Nishioka et al [1980] favor this

```
PSEUDO 30.5                                                          CCTAACTTCTTCCCAAACACCAGATGTTGGAG
ADULT ALPHA                                                         ATC CTGGAGA AT GTA GGG TAAGAAAGT
                                                                    -231
                                                                    -+---------+---------+---------+

PSEUDO 30.5   GCAGAGGAATGAGCAAGAAGCATCCGGGCAACTGATGAGGATTCCTTGCATCCAAGGAAGGCAGATCCCCAACCCAGGACTCTTGAGGGCTCTTAAAAGC
ADULT ALPHA   TGTCC GGCA CTG T   G  T CCTGC  CT G G G A GCA AAC CAG CCCAG AT TC GGGG CCTAACAAGT T ACT   TAGAGC  GCA
              -200                          -150
              +---------+---------+---------+---------+---------+

PSEUDO 30.5   CTGGCTGGGCAGAGCACAGGCCAGCCAATGAGGAG•CACTCAAAGGGCGTGTCTCAAAGAGCTGTAGGCATAGAAGTGCTGCTTATTTCAGGTCCAACAC
ADULT ALPHA   AAA CA C AT AGT ACTGCT   G GC TG TCCA CCT CCT GAGGA  GCCCTTG AG       T      A  GC G            G
              -100                          -50                                               -1
              7mGppp
               \

PSEUDO 30.5   ACTTCTGATTCTGATAGACTCAGGAAGACACCATGGTGCTCTCTGCGGAAGACAAAGCCGACGTCAGAACTGTCTGGGACAAGAGTGCAGGTCGTG•••C
ADULT ALPHA                            C                 A              G     AG A A AGG  C    GG   T GT C A  GTG
              1                               50
              ---------+---------+---------+

PSEUDO 30.5   TGACTCTGGCGCGGAAGCCAAAGGAAG••••••••//•••••••••GATGTTCGAGAGCATCTCCCC•••ACCAAGACCTACATCCCTCACTTTGAT
ADULT ALPHA        A A   T    CTG A  GTGAGAACAG// CTTCTCCCAG       T CT  ••  ACC              T  T
              100                                                     150
              --+---------+---------+---------+---------+

PSEUDO 30.5   GTAGGCCAGTGCTCGCCCCCAGGTCCAAGGTAATTGATGCACGG•••••••••GCTGATGC•••••••••••TGCAAACC••••••ATGGCCTGCCTG
ADULT ALPHA       A  CG    TG•     A G    •••••••  CAAGAAGGTC  C    GCTGGCCAGTGC  GG   ACCTCG  A      C
                                          200
              --------+---------+---------+          --------+        -----+--

PSEUDO 30.5   GTGCCCTATTGGCTCTGAGCAACCTGCACTCCCACAAGCTTCATGTTGGTTCCCGGACAACTTCAAG•••••••••//•••••••••CTCCTGAGCCAC
ADULT ALPHA         T G CT       G        TG     .    G G  A   CTG     GTATGCGCTG// GTCTCCGCAG
              250                          300
              --------+---------+---------+

PSEUDO 30.5   TGCTTGCTGGTGACCTCGGCTAG•CACCACCCTGCTGATTTCACCCCTCCCCTCGTGGTGCATGCCTCTCTTGACAAATTCCTTGCCTCTGT••••ACTG
ADULT ALPHA        C                 C              •••••• C A        G                                 GAGC  C
                                     350                              400
              ---------+---------+---------+

PSEUDO 30.5   TACTGACCTCCAAGTACTGATAATCTGCCTCCTGCCAGCCTTGCCTTCTGGCTAGACCCTTCTTCCCTCCCCTTGGATTCGTACCTCTTTGTCTTTGAATA
ADULT ALPHA   G                  CTG   T  GG G         C TG          CC CTT  CCTT  G
                           ---------+---------+          450                          500

PSEUDO 30.5   AAGCCCTGGGCCGGGGGGATCATTCTTGGGTCTGCA•CTGATGGAATAAAAGCTTATTTGAGTTCATGCTGATGGGCAGATGGAAGGTCTGCATTTTATGG
ADULT ALPHA       TGA TAG AA A G C G A  CC GGTTCT    CGTCTGC  G TG CA GTTTAGTG  GG  CCGCAGCTG  TTTG CATGGGGC GTA
              550                          600

PSEUDO 30.5   TATGGAACAGAAGACAGAGATGCTTAAATGAGGAGCAGCTTTCACCCTGGCACCGC
ADULT ALPHA   A GAC  GGTTCAGAGC AAAAGC TAATTG AT  C A    A ACA   ATA
              650                          699
              --+---------+---------+
```

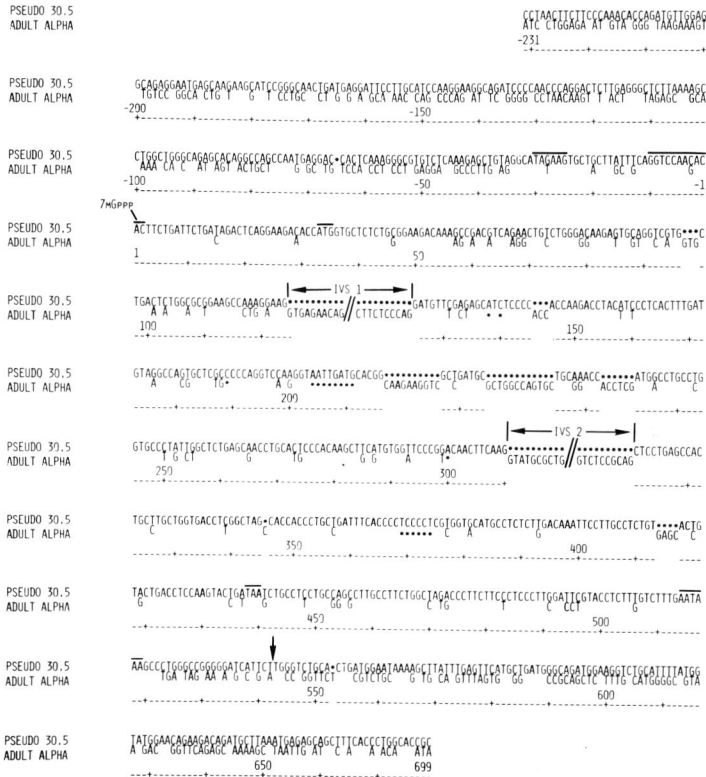

Fig. 7. A comparison of the sequences of a mouse pseudogene, ψα30.5 [Vanin et al, 1980], and adult alpha globin gene [Nishioka and Leder, 1979]. The nucleotide sequence of ψα30.5 (upper lines) is compared to the sequence of a productive adult alpha globin gene (lower lines). The sequence of ψα30.5 is listed in full. Where the sequence of the adult alpha globin gene differs from ψα30.5, the differing bases are shown; where they are identical, only the sequence of ψα30.5 is shown. Asterisks indicate that no bases are present at the indicated positions. The numbering system on the counting scale refers to ψα30.5, with number 1 corresponding to the nucleotide homologous to the first nucleotide of α globin mRNA. The intervening sequences, IVS 1 and IVS 2 of the adult α globin gene are shown. 7mGppp shows the position of the cap of adult α globin mRNA. Potentially important signals are overlined. The vertical arrow after nucleotide 542 shows the end of the adult α globin mRNA.

possibility. Whichever evolutionary scheme is correct, there remains the even more interesting question of whether pseudogenes have a function.

We have performed experiments to determine whether or not ψα30.5 is transcribed in mouse fetal liver—a tissue that is synthesizing large amounts of normal alpha globin polypeptide—and have been unable to find any evidence that ψα30.5 is transcribed in this tissue, using tests readily able to detect the

```
       1..........5.............10.............15...............21/23........26
(Met)ValLeuSerAlaGluAspLysAlaAspValArgThr ValTrpAspLysSerAlaGlyArgAlaAspSerGlyAla


     27.....................35
     GluAlaLysGlyArgMetPheGlySerIleSerProProArgProThrSerLeuThrLeuMet * AlaSerAlaArg


        54........57
     ProGlnValGlnGlyAsn * CysThrGly * CysCysLysProTrpProAlaTrpCysProIleGlySerGluGln


     ProAlaLeuProGlnAlaSerCysGlySerArgThrThrSerSerSer * AlaThrAlaCysTrp * ProArgLeu


                                                         134...136
     AlaProProCys * PheTyrProSerProArgGlyAlaCysLeuSer * GlnIleProCysLeuCysThrValLeu


     137......140
     ThrSerLysTyr *   *
```

Fig. 8. The amino acid sequence corresponding to the nucleotide sequence of $\psi\alpha30.5$ [from Vanin et al, 1980]. The amino acid sequence is that which would be obtained by translating $\psi\alpha30.5$ from the same initiator codon as the productive α globin gene. The numbered scales show amino acid residue numbers in α globins. In frame α globin amino acids are underlined. Amino acids in a different reading frame from α globin, or that have never been seen in an α globin, are italicized. Asterisks indicate chain terminator codons.

nuclear precursors of globin mRNA. This means that the pseudogene is either not being transcribed at all, or it is being transcribed at too low a level for us to detect, or its transcript is being degraded too rapidly for us to detect the steady-state level, or some combination of these factors is taking place.

Does this mean that the pseudogene $\psi\alpha30.5$ is a "dead" gene, or could it have a function? We have no direct answer to this question at present, but a good case can be made for the importance of pseudogenes in general by noting their widespread occurrence in other species. Maniatis and his collaborators [Lacy et al, 1979; Hardison et al, 1979; Fritsch et al, 1980; Lauer et al, 1980] have isolated a number of globin-related DNA sequences that cannot be identified with known polypeptides. In rabbits the gene β_2, located between the fetal and adult β globin genes, is a pseudogene in that it could not code for a functional globin polypeptide [Lacy and Maniatis, personal communication]. In humans there is a pseudogene, $\psi\alpha_1$, between the adult and fetal alpha globin genes which could not code for a functional globin polypeptide [Proudfoot and Maniatis, personal communication]. Both the rabbit pseudogene, $\psi\beta_2$, and the human pseudogene, $\psi\alpha_1$, have intervening sequences, but their DNA sequences are

Alu Family	CXACTCGGGAGGCTGAGGCAGGAGAATCGCATGAACC
BLUR 8 Clone	CCTGGAATCCCAGCTACTTAGGAGGCTCAGACAGAAGAATCCCTTAAACC
CHO Clone	TTTGCCCCAGXTACTCCAGAGGCAGAGGCCGCCACATTATGT
$^G\chi$ Gene	CCTGTAATCCCAGCTACTTGGGAGGCTGAGGAAGGAGAATTGCTTGAACC
Pseudogene	TGGGGAACTTCAGGT•CTTGGCAGGCTGAGGCAGGAGGATTGCTTGCGAG

<u>CONSENSUS SEQUENCE</u>   CCTG AA CCCAGCTACT GGAGGCTGAGGCAG AG AT C TGAACC

Alu I
SITE

41/50   ≤.1 DIFFERENCE

15/50 NO DIFFERENCES

Fig. 9.   A comparison of several members of the "*Alu* family" of repeated DNA. Five nucleotide sequences are compared from: the human "*Alu* family" sequence [Jelinek et al, 1980]; a clone of one member of the human *Alu* family, BLUR8 [Jelinek et al, 1980]; a Chinese hamster cell (CHO) cloned genomic DNA fragment [Jelinek et al, 1980]; a region on the 5' side of the human $^G\gamma$ fetal globin gene (our unpublished data); a region on the 5' side of the mouse pseudogene, $\psi\alpha$30.5 (our unpublished data). A consensus sequence is given below where not more than one nucleotide in the compared sequences is different (small capitals), or where there are no differences (large capitals). The *Alu* I site, AGCT, from which the family name is derived is shown.

abnormal in ways that would prevent the intervening sequences from being spliced out of the RNA in the usual manner [Lacy, Proudfoot, and Maniatis, personal communication]. Thus all three sequenced pseudogenes share the same basic faults: they could not code for functional globins, and they have abnormal or absent intervening sequences. When we add to this the fact that there are similar DNA sequences, which look as if they will prove to be pseudogenes, between the adult and prenatal β globin genes in both mouse [Jahn et al, 1980] and human [Fritsch et al, 1980], then a pattern begins to emerge. It looks as if there are pseudogenes between the prenatal and adult globin genes in all species that have been studied at the level required to characterize pseudogenes. We consequently think that it is unlikely that they are "dead" genes without function.

If they are not "dead" genes, what could their function be? An attractive possibility is that they could act to modulate the expression of the productive genes. There is considerable evidence indicating that globin genes may be "open" or "closed" at the chromatin level in the way originally suggested by Weintraub's pioneering work [Weintraub and Groudine, 1976]. Although the open config-uration is a necessary condition for a productive gene, it does not appear to be a sufficient condition. Several investigators have found that globin genes can be in the open configuration even though the corresponding globins are not being synthesized [Young et al, 1978; Stalder et al, 1980]. Consequently we are considering the possibility that production versus nonproduction of globins may be controlled by which gene within an open region is selected for transcription. If a pseudogene is selected for transcription within an open region then the transcript is diverted into a degradative pathway and no globin product is formed. If one of the normal genes is selected, then the nuclear transcript can be processed and transported to yield a cytoplasmic mRNA on which productive translation can occur. Thus the pseudogenes, on this hypothesis, may be acting as "diverting" genes which modulate the expression of the open globin regions by diverting the transcriptional machinery into a nonproductive pathway. There are partial precedents for the thought that mechanisms exist for selecting one of a number of products from a single transcriptional unit. In the adenovirus type 2 system different portions of the late transcript are used in different mRNAs [Nevins and Darnell, 1978]. In immunoglobulin heavy chains, the secreted and membrane forms of the μ-chain are probably the result of using different portions of the transcript from one region of DNA [Early et al, 1980]. Whether or not the "diverting" gene hypothesis proves to be correct, the finding of these pseudogenes raises many questions regarding their possible function in the globin gene clusters.

VIII.   REPEATED SEQUENCES IN THE β GLOBIN GENE CLUSTER

There is one more feature in the globin gene cluster that I want to describe. It concerns the distribution within these regions of some repeated DNA sequences. Three independent sets of observations have pointed to their existence in the β gene cluster (I shall confine my remarks on this topic to the β gene cluster). The first observation [Duncan et al, 1979] was the location of two places in the β gene cluster that could be transcribed in vitro with RNA polymerase III. (This is the polymerase that transcribes the DNA for 5S RNA, tRNA and some other small molecular weight RNAs; it is not thought to be required for the transcription of protein structural genes.) These two pol III sites were located 5' to the fetal $^G\gamma$ gene, and 5' to the adult δ gene. The transcript from the site 5' to $^G\gamma$ would hybridize to the site 5' to the adult δ gene, and also to another site 3' to the whole β globin cluster. The second observation was made by Fritsch et al [1980], who described the presence in the β globin gene cluster of six regions of DNA, all able to hybridize to a probe made from one of them (see Fig. 1). The six regions included all three regions discovered by Duncan et al [1979]. The third observation, described in a paper by Jelinek et al [1980], hints at the possible nature of these repeated sequences. The paper presents a comparison of the DNA sequences from a number of sources. Included in these sequences is that derived from a family of repeated DNA, which is interspersed throughout the human and hamster genome approximately 300,000 times. This family of interspersed repeated DNA is referred to as the *Alu* family, because many of its members are cut with the restriction enzyme *Alu* I. The pol III site 5′ to the ᴳγ gene proves to be a member of this *Alu* family on the basis of its DNA sequence. This suggests, although it does not prove, that the other five repeated sequences in the β cluster are also members of this family.

Why is this possibility exciting? The answer is that Jelinek et al [1980] showed that the sequence common to the *Alu* family also has much in common with the known origins of replication of several DNA viruses. There is consequently a real possibility that for the first time we will soon have some knowledge of the relationships between the location of origins of DNA replication and the location of structural genes that are expressed at different times during development. Such a correlation would be of great interest, since many developmental biologists suspect that determination and differentiation are intimately related to DNA replication.

Finally, let me show you that the *Alu* family sequence occurs also near the pseudogene ψα30.5. Figure 9 compares five DNA sequences: the sequence reported by Jelinek et al [1980] for the *Alu* family in general, the sequence of two cloned members of the *Alu* family [also from Jelinek et al, 1980], our sequence of the member of the *Alu* family which is 5′ to the ᴳγ gene, and our sequence from (the 5′ side of) the pseudogene ψα30.5. The fidelity of preservation of these sequences is remarkable in that, of a total of 50 nucleotides, there are 41 positions where there is not more than one difference among the five sequences, and at 15 of these positions there are no differences. The degree of preservation of a nucleotide sequence in DNAs from so many varied sources points to its importance.

IX. CONCLUSION

I must conclude, and as I do so I hope I am leaving you with the picture of a developmentally complex locus which, as a result of DNA cloning and sequencing, is beginning to shows signs of order. The *evolution* of the globin gene clusters is beginning to be understood. An unexpected class of genes has been demonstrated, the *pseudogenes,* which may be important in modulating the expression of these globin gene clusters. Finally, the location within the clusters of several repeats that may prove to be *origins of replication* is now known.

ACKNOWLEDGMENTS

This is paper number 2438 from the Laboratory of Genetics at the University of Wisconsin. The work was supported by NIH grants GM20069 and AM20120.

REFERENCES

Blattner FR, Blechl AE, Denniston-Thompson K, Faber HE, Richards JE, Slightom JL, Tucker PW, Smithies O (1978). Science 202:1279–1284.

Dayhoff MO (1972). In "Atlas of Protein Sequence and Structure." (MO Dayhoff, ed.), vol 5, pp 17–30. National Biomedical Research Foundation, Washington, D.C.

Duncan C, Biro PA, Choudary PV, Elder JT, Wang RRC, Forget BG, DeRiel J, Weissman SM (1979). Proc Natl Acad Sci USA 76:5095–5099.

Early P, Rogers J, Davis M, Calame K, Bond M, Wall R, Hood L (1980). Cell 20:313–320.

Efstratiadis A, Posakony JW, Maniatis T, Lawn RM, O'Connell C, Spritz RA, DeRiel JK, Forget B, Weissman SM, Slightom JL, Blechl AE, Smithies O, Baralle FE, Shoulders CC, Proudfoot NJ (1980). Cell 21:653–668.

Fritsch EF, Lawn RM, Maniatis T (1980). Cell 19:959–972.

Hamer DH, Smith KD, Boyer SH, Leder P (1979). Cell 17:725–735.

Hardison RC, Butler III ET, Lacy E, Maniatis T, Rosenthal N, Efstratiadis A (1979). Cell 18:1285–1297.

Jahn CL, Hutchison III CA, Phillips SJ, Weaver S, Haigwood NL, Voliva CF, Edgell MH (1980). Cell 21:159–168.

Jelinek WR, Toomey TP, Leinwand L, Duncan CH, Biro PA, Choudary PV, Weissman SM, Rubin CM, Houck CM, Deininger PL, Schmid CW (1980). Proc Natl Acad Sci USA 77:1398–1402.

Lacy E, Hardison RC, Quon D, Maniatis T (1979). Cell 18:1273–1283.

Lauer J, Shen CKJ, Maniatis T (1980). Cell 20:119–130.

Leder A, Miller HI, Hamer DH, Seidman JG, Norman B, Sullivan M, Leder P (1978). Proc Natl Acad Sci USA 75:6187–6191.

Nevins JR, Darnell JE (1978). Cell 15:1477–1493.

Nishioka Y, Leder P (1979). Cell 18:875–882.

Nishioka Y, Leder A, Leder P (1980). Proc Natl Acad Sci USA 77:2806–2809.

Radding CM (1978). Annu Rev Biochem 47:847–880.

Slightom JL, Blechl AE, Smithies O (1980). Cell 21:627–638.

Smithies O (1964). Cold Spring Harbor Symp Quant Biol 29:309–319.

Smithies O, Blechl AE, Denniston-Thompson K, Newell N, Richards JE, Slightom JL, Tucker PW, Blattner FR (1978). Science 202:1284–1289.

Stalder J, Groudine M, Dodgson JB, Engel JD, Weintraub H (1980). Cell 19:973–980.

Vanin EF, Goldberg GI, Tucker PW, Smithies O (1980). Nature 286:222–226.

Weintraub H, Groudine M (1976). Science 193:848–856.

Young NS, Benz Jr EJ, Kantor JA, Kretschmer P, Nienhuis AW (1978). Proc Natl Acad Sci USA 75:5884–5888.

On the Control and Regulation of Size and Morphogenesis in Mammalian Embryos

M.H.L. Snow, P.P.L. Tam, and Anne McLaren

MRC Mammalian Development Unit, Wolfson House, 4 Stephenson Way, London NW1 2HE, England

I. INTRODUCTION

When we buy a kitten, a puppy, a calf, a baby hamster, or an elephant we have at least some idea of how big it will become. We do not expect a Pekinese puppy to grow as large as a Great Dane or a Shetland pony to become as big as a Shire horse. The very existence of these strains with large differences in size, selected by man, indicates a genetic component in growth and regulation of size. In mammals final body size is determined and generally achieved long before death. Mammals share this property with birds and many, but not all, lower vertebrates. Some fish, amphibians, and reptiles, for instance, grow throughout life [Goss, 1974].

In mammals, furthermore, given adequate diet and health, the progression to final size—ie, growth—is sufficiently characteristic of a species or a strain

Levels of Genetic Control in Development, pages 201–217

for tables of normal values to be constructed that will give, within fairly narrow limits, the length, height, width, or weight an individual should be at a given age. For some well-studied species (eg, man), these tables of normal values give details down to the level of the dimensions of individual bones and organs.

The size any organism finally attains is the result of controlled growth, and the regulation of size is essentially a matter of the rate and duration of growth. There is not a consistent relationship between these parameters among mammals, some achieving their final size quickly by rapid growth, others more slowly and at a more leisurely growth rate. This fact was recognized by Minot [1891], who estimated that from conception to maturity guinea pigs grew at an average rate of 1.8 gm/day, rabbits at 6.3 gm/day, and humans at 6.79 gm/day. Thus rabbits are bigger than guinea pigs because they grow faster, but humans are bigger than rabbits because they grow for a longer time. This simple generalization seems also to apply to the period from conception to birth. As McCance and Widdowson [1978] point out, the newborn hippopotamus is larger than a newborn man, despite a shorter gestation time, because it grows more rapidly in utero. Similarly, the newborn elephant is about the same weight as a newborn horse but achieves this weight after spending twice as long in utero. Similar associations are observed during the early stages of pregnancy; hamsters, rabbits, rats, and mice reach a weight of 1 gm in about two weeks, pigs and cats in four to five weeks, baboons and man in about eight weeks [Hendrickx, 1971; Adolph, 1972; McLaren, 1976a].

Normal tables of embryonic development also give detailed morphological descriptions characteristic of embryos of specific ages. Consequently there has arisen a belief that growth and morphogenesis must be closely coordinated [see review by Snow, 1978]. Certainly in most embryos the relationships in size between the component organs are remarkably regular, and any deviation in size, up or down, of one particular organ is instantly recognized by the embryologist and classified as abnormal.

In this paper we have gathered together some data that have a bearing on the regulation of growth and size in embryos. The data are presented in three sections. The first deals with the genetic and physiological constraints acting between and within the uterine environment and fetus. The second describes the outcome of artificial manipulation of embryonic size prior to implantation, and the last section comments on our recent observations on compensatory growth following extensive cell death in post-implantation embryos.

II. MATERNAL ENVIRONMENT VS EMBRYONIC GENOTYPE

Assuming that the embryo is normal and healthy, the uterine environment can clearly influence embryonic growth by limiting nutrition. In man several maternal factors are known to impede fetal nutrition and to result in small-for-dates babies. They include such diverse factors as pregnancy at high altitude, multiple pregnancy, maternal heart disease, and anaemia [Ounsted and Ounsted, 1973; Gruenwald, 1974]. Maternal undernutrition and heat stress are known also to result in low birth weight and high perinatal mortality in lambs [Alexander, 1974].

Nevertheless we shall assume that nutrition is not limiting and pose the question: Is there a Parkinson's law of pregnancy? C. Northcote Parkinson [1958], in a satirical analysis of the growth of bureaucracy, postulated that work and/or the workforce expand to fill the time and space available. Put in fetal growth terms, does fetal size reflect uterine space, a function of maternal size? The classical experiments addressing this question are those of Walton and Hammond [1938], who studied the offspring of matings between Shire horses and Shetland ponies. There is approximately a fourfold difference in weight between the mares; reciprocal crosses, achieved by artificial insemination, provided embryos of substantially the same genotype growing in uteri of vastly different sizes. The hybrid foals born to Shetland mares were comparable in size to pure-bred Shetlands, while the foals born to Shire mares were considerably larger, approaching but not quite equaling the weight of pure-bred Shire foals. Both types of foal were of normal proportions. The clear implication is that fetal size does reflect maternal size. Walton and Hammond's study went further to follow the growth of the foals from these different dams. Only after weaning, when the foals were kept under the same nutritive conditions, did the effect of genotype emerge. The hybrid foals from Shire mares grew more slowly than pure Shires, and those from Shetland mares grew more rapidly than pure Shetlands. However, the advantages conferred on the hybrid by growth in the Shire uterus were maintained and still reflected in considerable size differences at three years of age.

With respect to size at birth, these results have been substantiated in cattle [Joubert and Hammond, 1954; Dickinson, 1960] and sheep [Hunter, 1956]. In his work on sheep, Hunter eliminated the possible role of cytoplasmic inheritance through the egg by doing reciprocal transfers of fertilized eggs between ewes of the different strains. These results are summarized in Table I.

The general conclusion from these studies, all done on species that usually are monotocous, seems to be that maternal uterine size has a greater influence on fetal growth than embryonic genotype. If the polytocous mouse is considered, the resulting conclusion is different.

Mice have been selected for high and low body weight at six weeks of age [Elliott et al, 1968; White et al, 1968; Falconer, 1955, 1973], which has resulted in sublines differing up to twofold in adult size. The difference in growth of the large and small sublines of the Q strain mice used by Falconer is first detectable in embryos at the beginning of organogenesis, at about eight days post coitum [I. Gauld, quoted by Blakely, 1979]. Although there is a negative correlation

TABLE I
Fetal Weight Variation According to Uterine Environment

Species	Strain	Birth weight (lbs) from		Reference
		Large dam	Small dam	
Horse	Shire	156	—	Walton & Hammond [1938]
	Shetland	—	43	
	hybrid	109	39	
Cattle	S. Devon	95	—	Joubert & Hammond [1954]
	Dexter	—	52	
	hybrid	72	58	
Sheep	Border Leicester	14	11	Hunter [1956]
	Welsh Mountain	10	8	

TABLE II
Mouse Fetal Weight (gm) at Birth* or 16 Days of Gestation**

Reference	Embryo genotype	Uterine genotype		
		Large	unselected	Small
Brumby [1960]	Large	—	1.74(31)	1.36(32)
	Small	1.31(37)	1.43(58)	—
Moore et al* [1970]	Large	1.67(49)	—	1.57(79)
	Small	1.43(75)	—	1.34(88)
Aitken et al** [1977]	Large	0.639(57)	0.624(54)	—
	Small	—	0.524(67)	0.554(63)

between litter size and fetal weight in mice [McLaren, 1965; Falconer, 1965; Roberts, 1966], the fact that genetically large mice come from significantly larger litters than small mice [Brumby, 1960; Bowman and McLaren, 1970] indicates that this parameter is not important in determining the difference between the lines.

The technique of embryo transfer has been used to analyze the respective effects of fetal genotype and uterine environment. The data are summarised in Table II. Brumby [1960] found an effect of both maternal environment and fetal genotype, with the major influence exerted by the uterus. Moore et al [1970] similarly found effects of both factors in mice of different origin, but in their study the most important parameter was fetal genotype. These two early studies differ in one important aspect, which makes comparison difficult. Brumby [1960] used asynchronous transfers—ie, embryos taken from a female three and one-half days post coitum were put into the uterus of a foster mother who was two and one-half days pseudo-pregnant—but failed to control for the effect that an additional 24 hours in utero may have on fetal weight. Since Aitken et al [1977] in similar experiments have shown that an extra day in utero can elevate "small" fetal size to a value in excess of that found in normal matings of "large" animals, it is not possible to evaluate accurately the respective roles of uterus and fetal genotype in Brumby's study. Aitken et al [1977], using mice of Falconer's Q strain, made synchronous transfers of large or small embryos into foster mothers of the parent, unselected strain, so that both fetal genotypes were developing in identical uterine environments. In these conditions fetal weight on day 16 of gestation reflects only fetal genotype, there being no apparent influence of the uterus either to depress the growth of large embryos or to enhance that of small embryos.

Conclusions

Both uterine environment and fetal genotype can influence fetal growth. In large monotocous species maternal size seems to exert the major influence, but in the polytocous mouse, fetal genotype exerts the greater control.

III. MANIPULATIONS OF PRE-IMPLANTATION EMBRYOS

The development of in vitro culture techniques applicable to pre-implantation stage mouse embryos made possible manipulations that either decrease or increase cell number in the embryo. Reductions in cell number are brought about either by destruction of cells [Tarkowski, 1959a, b], separation of blastomeres [Mullen et al, 1970], or suppression of cell division [Snow, 1973]. Increases are achieved by adding cells, either using the technique of embryo aggregation [Tarkowski, 1961] or by injection of cells into the blastocyst [Gardner, 1968]. These various methods have been adapted and applied to rabbits [Moore et al, 1968; Moustafa, 1974; Gardner and Munro, 1974], rats [Mayer and Fritz, 1974] and sheep [Tucker et al, 1974; Willadsen, 1979].

With the exception of cell number reduction by suppression of cleavage, which results also in doubling of chromosome number to yield tetraploids and results in abnormal development [Snow, 1975, 1976b; Tarkowski et al, 1977], these techniques permit the production of normal, viable offspring that are born at the expected time and that are of normal size. Hence small embryos regulate upwards and large ones regulate downwards in size. The extent to which regulation is possible can be quite extensive; Moore et al [1968] produced viable rabbits from 11% of transfers of one cell from the eight-cell stage. Willadsen [1979] produced identical twin sheep from two-cell embryos, and more recently has made identical quadruplets from four-cell embryos [personal communication]; normal chimeric mice have been made from aggregates of three embryos [Markert and Petters, 1978] and can routinely be produced from two [see review by McLaren, 1976b].

When and how does this regulation take place? There is no clear answer as to how, but the timing of regulation is known at least in the mouse. The downward regulation in size required for aggregate chimeras to achieve normal size occurs shortly after implantation but before primitive streak formation and the establishment of overt bilateral symmetry. In their study of Q strain mice, Buehr and McLaren [1974] found that chimeric embryos are clearly larger than singles up to five and one-half days post coitum (about 30–36 hours after implantation), are not always distinguishable between five and one-half and six days, and invariably are the same size thereafter.

Upward regulation in size occurs at a different time. Tarkowski [1959b] followed the development through gestation of ½ embryos—ie, those developing from one cell of the two-cell stage—and found that 11/12 embryos examined between days six and ten of pregnancy were morphologically normal but about half the size of control embryos. Fetuses on the eleventh day were variable, but all appeared to be of normal size from the twelfth day onwards. No attempts were made to determine how the "catch-up" growth was achieved, whether it was by increased cell proliferation, increased cell size, increased extracellular material, or a combination of factors.

Although the mechanisms of regulation are poorly understood, the different times at which upward and downward changes occur may reflect the changing conditions of maternal/fetal relationships. In the few days immediately following implantation the primary trophoblast cells invade and erode the uterine decidual tissue surrounding the developing conceptus, creating lacunae that become gorged with maternal blood. The final breakdown of tissue resulting in maternal blood flooding the spaces immediately adjacent to the developing embryo occurs between six and six and one-half days [Theiler, 1972]. Since embryonic nutrition in these early stages must rely on diffusion of material across the decidual tissue, it seems plausible that availability of nutrients may be a limiting factor determining growth rate. Indeed, at about this stage of development in normal embryos there seems to be a lowering of mitotic index, which is then followed by a large acceleration in growth rate coinciding not only with primitive streak formation in the embryo but also with decidual breakdown in the uterus [Snow, 1976a, 1977]. Similar nutritional changes may be involved in the timing of the upward regulation in size. The chorio-allantoic placenta is formed at about nine to nine and one-half days in the mouse, but complete formation of the labyrinth in which exchange between maternal and fetal circulation is most efficient may not occur until some 24 hours later [Friedrich, 1964]. It may therefore not be coincidence that the "catch-up" growth immediately follows this period.

The downward regulation in size of chimeric embryos provides little information of use in understanding mechanisms for controlling size and morphogenesis. Since it could result from nutritional constraints imposed by the maternal environment, there is no compelling reason to invoke mechanisms intrinsic to the embryo. The same cannot be said for the upward size regulation. Two important observations emerge from these studies. First, morphogenesis is not affected by experimental reduction of fetal size and must therefore be regarded as under independent control, a subject that is discussed in greater detail later. Second, irrespective of the possibility that the accelerated growth is facilitated by improved nutrition, the fetus must recognize its own smallness and set in motion a novel internal program to correct the deficit, since there is no such spurt in growth rate in normal embryos of the same age [MacDowell et al, 1927; Goedbloed, 1972].

The fact that the fetus maintains normal body proportions during this disturbed growth pattern suggests that the growth-regulating factor(s), whatever it may be, acts systemically [see also Blakely, 1979]. Is this hypothetical factor of local origin?—ie, is there a growth-controlling center in the fetus capable of monitoring size and altering growth rate if necessary? Studies of chimeras between large and small strains of mice provide a partial answer to this question [Roberts et al, 1976; Falconer et al, 1978]. These analyses were carried out on adult chimeras, but since the difference in growth rate in the two component strains is apparent in early embryogenesis (see above), it seems reasonable to presume that growth control in the chimera will also operate during gestation. The initial study, in which 31 overt coat-color chimeras were used, demonstrated a positive correlation between body weight and the proportion of the large genotype represented in the coat colors. Roberts et al [1976] argued that it was unlikely that melanocytes themselves were the determinants of body size but that the observed correlation reflected similarities in the cellular composition of the coat and whatever organs were involved in growth regulation. In their subsequent study they made chimeras in which the component strains differed with respect to the isozymes of glucose phosphate isomerase, so that by using electrophoretic analysis the study could be extended to include brain, spinal cord, pituitary, liver, kidney, spleen, and blood [Falconer et al, 1978]. Several important points emerge: 1) Cell selection leading to a preponderance of "large" genotype did not occur. 2) Organs within individual mice showed significant differences in their cellular composition, so that a growth-controlling organ could have been identified had one existed. 3) The overall data suggest that no one organ has a controlling effect on body weight. 4) Body weight correlates well with the mean cell proportion of *all* organs. Furthermore, the cell proportions of the organs studied accounted for all the variance expected to arise from genetic differences between the constituent strains; there was no residual variance to suggest that some other organ or tissue might control growth. However, as Falconer et al point out, this could occur if the nine organs (coat-color variegation included) together gave an accurate estimate of the cell proportions of the unknown growth-controlling organ. 5) Cell proportions in an organ did not convincingly correlate with the weight of the organ itself. Organ size, as in normal, non-chimeric animals, seemed to reflect whole body size. Recent unpublished data suggest that liver and kidney size may possibly vary according to cellular constitution, but confirm that no such correlation is present in other organs [Falconer, personal communication].

Conclusions

Embryos of widely differing species apparently are capable of assessing their size and of comparing it to a notional "target" size [Tanner, 1963] appropriate for their age. If small, then the embryo can accelerate its growth to bring size back to normal. Morphogenesis seems unaffected by these variations in size or growth rate, giving the impression that the factors regulating growth either act systemically or are very closely coordinated at organ level throughout the embryo. A study of chimeras between large and small mice has failed to identify a growth-controlling organ and would seem to suggest that growth (and its control) is a function of genotype and acts at the level of the cell.

IV. COMPENSATORY GROWTH IN THE POST-IMPLANTATION EMBRYO

It has been shown recently [Snow and Tam, 1979] that the drug mitomycin C (MMC) causes random cell death in primitive streak stage mouse embryos (6.5–7.5 days post coitum) in utero and can reduce cell numbers to 10–15% of normal values at 7.5 days post coitum. Morphogenesis is not immediately disturbed, and all embryos complete gastrulation normally, producing a very small neural plate stage embryo at eight days post coitum. Mitotic index in all tissues of these embryos is much higher than normal, showing that accelerated compensatory growth has commenced. Judged by axis length, embryos achieve normal size by about 12 days post coitum, but fetal weight is not normal until after 13.5 days post coitum (Table III).

During the 48 hours of development from eight days post coitum morphogenesis is disturbed in a rather complex manner, apparently to different degrees in different tissues or organ systems. Figure 1 shows outline drawings of embryos at various ages, illustrating the gross developmental differences between normal

TABLE III
Weights and Linear Dimensions of Normal (N) and MMC-treated Embryos

AGE (days)	Fetal weight, gm, (n)		Axis length, mm,[a] (n)	
	N	MMC	N	MMC
8.5	—	—	1.25 (41)	0.68 (29)
9.5	1.9 (11)	1.0 (12)	5.39 (46)	3.07 (58)
10.5	8.9 (22)	3.8 (28)	9.20 (41)	7.07 (55)
11.5	27.8 (23)	17.8 (32)	12.18 (27)	11.00 (14)
12.5	63.2 (20)	45.9 (24)	14.52 (18)	14.57 (22)
13.5	143.0 (42)	126.0 (42)	—	—
Birth	1419.0 (52)	1362.0 (27)	—	—

[a]Axis length was measured from the center of the otic capsule to the tail tip.

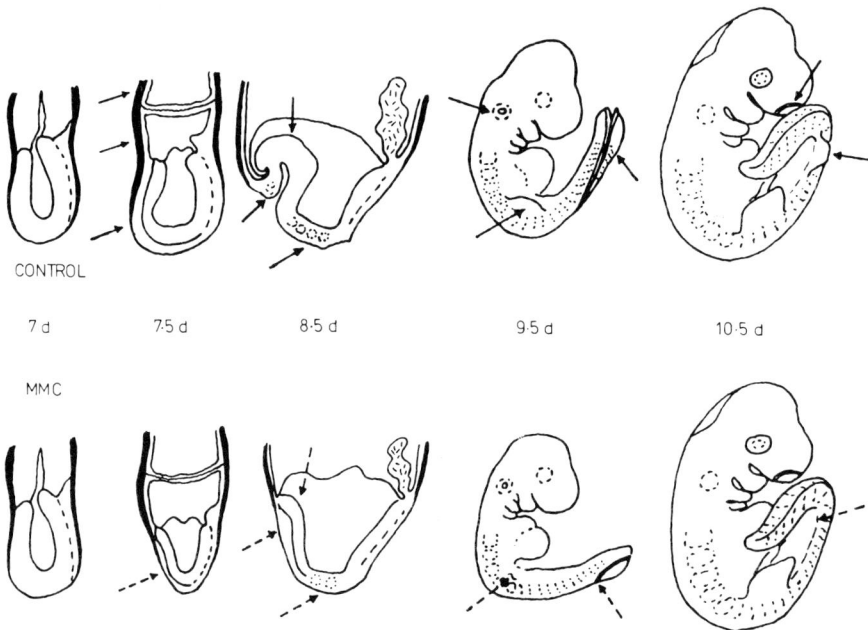

Fig. 1.   Outline drawings of embryos of different ages from control and mitomycin C-treated litters. The solid arrows indicate the diagnostic features of the normal embryos and the broken arrows the points in which the MMC embryos differ.

and MMC-treated embryos. The first obvious divergence from the normal course of development is the retardation in neural fold formation, which may be delayed some nine to ten hours. However, once the head folds are formed, flexure of the head follows more swiftly than in normal embryos, apparently occurring at the "correct" chronological age. Thereafter the gross developmental horizons of neural tube development are achieved "on time," although with a neural ectoderm of significantly reduced size in early stages. The developmental profile of mesodermal components (somites, limb buds, etc) follows a somewhat different timetable. The first somite is formed at the normal chronological age, the second one is a little delayed, and there is a progressively longer delay in formation of successive somites up to about ten days post coitum when, according to somite number, the MMC embryos seem about a half-day retarded. Thereafter the rate of somitogenesis in the MMC embryos is maintained, while that in normal embryos declines, so that the gap narrows and eventually closes. There is thus a transient discrepancy between the developmental status estimated by somite number and that suggested by neural tube morphology. The rest of this chapter will focus in more detail on the temporal mismatch between organ systems during the compensatory growth of MMC-damaged embryos.

A. Assessment of Embryonic Age

Because of the obvious "mismatch" between neural tube morphology and somite number in the MMC embryos, and because in any one litter the embryos show a range of somite numbers, it is not appropriate to equate developmen-

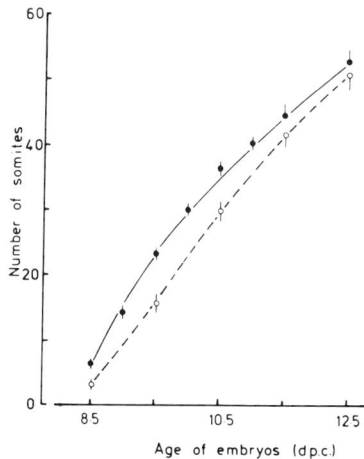

Fig. 2.   Somite number plotted against age. ● = normal; ○ = MMC-treated embryos.

tal status of any organ system to either somite number or litter age. Therefore, comparisons have been made on the basis of a notional developmental age, calculated as follows: Somite numbers of embryos in normal or MMC litters autopsied at half-day intervals are plotted against age post coitum. From this graph the "average" age of an embryo with a given somite number can be calculated. Figure 2 shows this plot for normal and MMC-treated embryos; the calculated best fit regression lines show that normally the first somite forms at 8.34 days post coitum, somite 15 at 9.00 days, somite 30 at 9.92 days, and somite 40 at 10.86 days post coitum. For MMC embryos the corresponding figures are 8.40, 9.38, 10.47, and 11.18 days post coitum, respectively. Thus the greatest retardation in somite formation is found at around the expected 30 somite stage, or ten days post coitum.

B. Morphogenesis of the Neural Tube

The gross developmental horizons measured were closure of the anterior neuropore, brain roof, otic capsule and posterior neuropore, and formation of the olfactory pits. Each event was scored with respect to the embryo somite number. The "age" of the embryo was then calculated from the graph shown in Figure 2. In normal embryos the morphogenesis of the neural tube, at least according to the parameters studied, is extremely well correlated with somite number and hence with developmental age. In MMC embryos the same morphogenetic events occur at different somite numbers but at the same developmental age (Table IV). Thus the development of somites and neural tube are not dependent upon, nor correlated with, one another during this compensatory growth.

C. Morphogenesis of Limb Buds

Table IV also includes data on the timing of limb-bud formation, and suggests that the fore- and hind-limbs may form under different controls. The forelimb appears about 14 hours later than normal in MMC embryos but at about the correct somite number. The hind-limb bud, however, appears at both an abnormal time and an abnormal somite number.

These data clearly indicate that hind-limb bud formation is independent of both somite status and neural tube status. Forelimb formation could be related to somitogenesis.

D. Germ Cell Population

Primordial germ cells (PGCs) are identifiable during organogenesis by their high alkaline phosphatase content. Figure 3 shows the numbers of PGCs in normal and MMC embryos between 8.5 and 13.5 days post coitum. Unlike

TABLE IV
Morphogenesis in Normal (N) and MMC-treated Mice

Developmental Feature	Embryos		Somite Numbers		Estimated Age (days)
	Type	Number	Range Covered	Timing of Event[a]	
Ant. neuropore closure	N	20	11–17	14–15	9.0
	MMC	19	5–11	7–9	8.9
Brain roof closure	N	20	11–17	15–16	9.0
	MMC	20	9–14	10–11	9.0
Otic capsule closure	N	17	24–29	26–27	9.7
	MMC	27	20–27	23–24	9.8
Post-neuropore closure	N	32	23–29	25–26	9.6
	MMC	18	22–27	25–26	9.9
Nasal pit formation	N	30	27–33	29–30	9.9
	MMC	30	24–30	26–27	10.0
Forelimb bud formation	N	21	17–22	19–20	9.2
	MMC	39	17–26	21–22	9.8
Hind-limb bud formation	N	34	27–34	31–32	10.0
	MMC	36	25–31	27–28	10.2

[a]Somite number below which it was never observed and above which it was always observed.

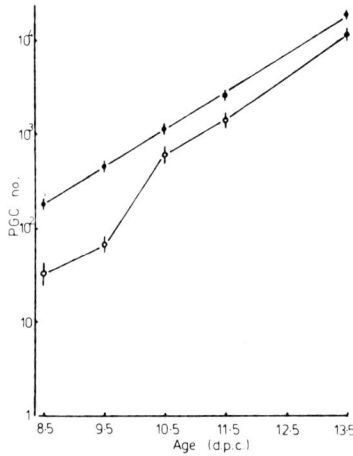

Fig. 3.   Increase in primordial germ cell number with age. ● = normal; ○ = MMC-treated embryos.

other organs, the germ cell population does not recover to normal size. The
genital ridges are colonized by about half the normal numbers of PGCs because
the compensatory proliferation seems to be restricted to the period between 9.5
and 10.5 days post coitum. During their normal proliferative phase the PGCs
migrate from the base of the allantois, via the hind gut and its mesentery, into
the genital ridges. Comparisons of the position in the migration path with age
or somite number indicate that the germ cell population is acting independently
of age or somite number and, hence, independently of neural tube or limb bud
development as well. Figure 4 shows the positions of PGCs according to litter
age, and Figure 5 according to somite number. The migration of PGCs can be
described as retarded or advanced, depending on the assessment criterion used.

E. Differentiation of Somites

Some time after its formation the somite undergoes a clear morphogenetic
change whereby the wall towards the interior of the embryo breaks down, and
cells, the sclerotome cells, disperse internally (Fig. 6). The time elapsed between
somite formation and sclerotome dispersion can be calculated from histological
sections of embryos of known somite number. Figure 7 plots this differentiation
time for somites 2 to 38 in normal and MMC-treated embryos. For somites 2–14,
differentiation is slower in MMC embryos, but somites 18–32 develop more
quickly than normal. Hence differentiation of somites does not follow a chronological time scale; consequently it seems to be independent of neural tube
development.

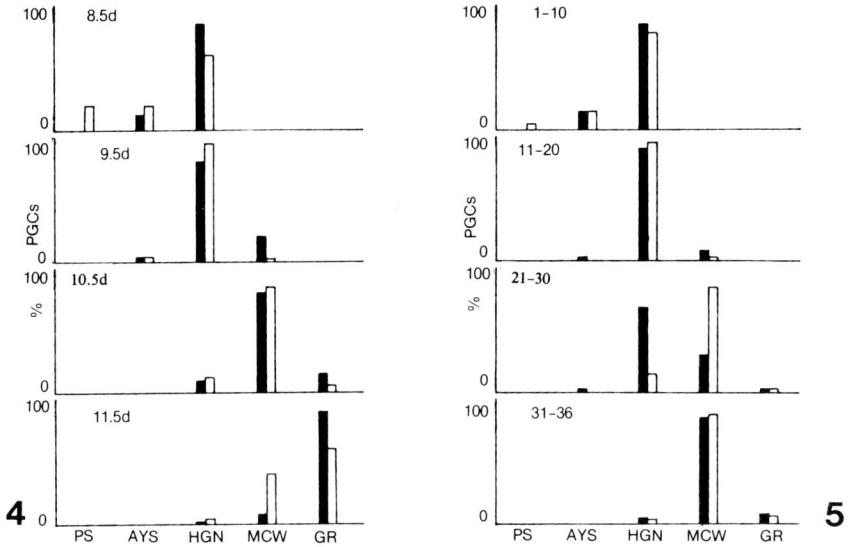

Fig. 4. (left) The position of PGCs in their migration path with respect to age. PS = primitive streak, AYS = allantois and yolk sac, HGN = hind-gut endoderm, MCW = mesentery and coelomic wall, GR = genital ridge, ■ = normal; □ = MMC.

Fig. 5. (right) As Figure 4, but plotted with respect to somite number.

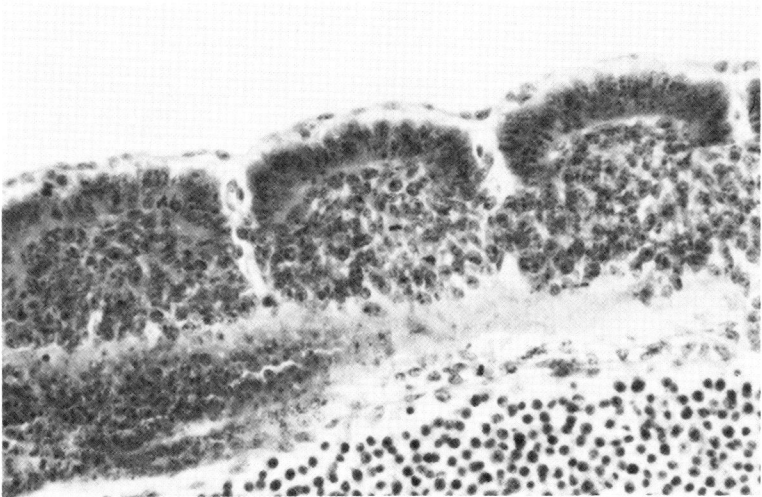

Fig. 6. Dispersal of sclerotome cells as a somite differentiates. The somite on the left is intact, with its internal surface (lower edge) intact, whereas the somite on the right is beginning to disintegrate and cells are starting to disperse internally.

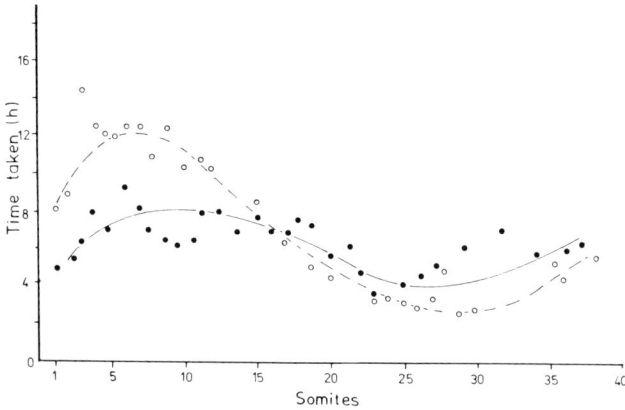

Fig. 7. The time taken between somite formation and the onset of sclerotome dispersal (see Fig. 6.) for somites 1–38. ● = normal; ○ = MMC.

V. CONCLUSIONS AND COMMENTS

The impression gained from the brief description of embryogenesis during compensatory growth is that coordination between various parts of the mouse embryo is not necessary for the formation of an anatomically normal fetus. Certainly at the gross level at which the analysis has been carried out, this seems to be true, but post-natal studies clearly demonstrate that all is not normal. Early post-natal mortality is extremely high, with over 60% of newborns dying within 14 days of birth. There is a high incidence of reduced fertility among those that survive weaning [Snow and Tam, 1979]. The processes underlying the reduced fertility have been identified [Tam and Snow, in press], but the cause of post-natal mortality is unclear. There is commonly a syndrome of runting and neurological disturbance that may be the outcome of the developmental mismatch between neural tube and somitic development, which is maximal when spinal ganglia and nerve outgrowths should be formed.

Nevertheless, little if anything seems to be wrong with many of the survivors. Since no embryo escapes the ravages of MMC in utero and none was encountered without developmental mismatch between organs, it must be concluded that embryogenesis and production of a "normal" functional animal can be achieved despite considerable interference in the apparently coordinated development of the various organ systems. It is concluded, therefore, that different tissues and organs are under intrinsic and autonomous control and that each has, to a greater or lesser extent, the ability to regulate its own rate of growth towards its target size.

A knowledge of that target size, however, need not reside individually in each organ. It remains possible that this property exists only in the most rapidly growing tissue—ie, the ectoderm—and that other tissues use this as a reference to monitor their own size and modify their own growth rate as best they can according to any discrepancy that arises.

REFERENCES

Adolph EF (1972). In "Nutrition and Development." (M Winick, ed.). pp 1–25. Wiley, New York.

Aitken RJ, Bowman P, Gauld IK (1977). J Embryol Exp Morph 37:59–64.

Alexander G (1974). In "Size at Birth." (K Elliot and J Knight, eds.), Ciba Found Symp. 27:215–245. Elsevier, Amsterdam.

Blakely A (1979). Genet Res Camb 34:77–85.

Bowman P, McLaren A (1970). Genet Res Camb 15:261–263.

Brumby PJ (1960). Heredity 14:1–18.

Buehr M, McLaren A (1974). J Embryol Exp Morphol 31:229–234.

Dickinson AG (1960). J Agr Sci 54:378–390.

Elliott DS, Legates JE, Ulberg LC (1968). J Reprod Fertil 17:9–18.

Falconer DS (1955). Cold Spring Harbor Symp Quant Biol 20:178–196.

Falconer DS (1965). Proc XIth Int Cong Genet 3:763–774.

Falconer DS (1973). Genet Res Camb 22:291–321.

Falconer DS, Gauld IK, Roberts RC (1978). In "Genetic Mosaics and Chimaeras in Mammals." (LB Russell, ed.), pp 39–49. Plenum, New York & London.

Friedrich F (1964). Z Anat Entwickl-Gesch 124:153–170.

Gardner RL (1968). Nature, 220:596–597.

Gardner RL, Munro AJ (1974). Nature 250:146–147.

Goedbloed JF (1972). Acta Anat 82:305–336.

Goss RJ (1974). Perspect Biol Med 17:485–494.

Gruenwald P (1974). In "Size at Birth." (K Elliot and J Knight, eds.), Ciba Found Symp 27:3–26. Elsevier, Amsterdam.

Hendrickx AG (1971). "Embryology of the Baboon." University of Chicago Press, Chicago and London.

Hunter GL (1956). J Agr Sci 48:36–60.

Jourbert DM, Hammond J (1954). Nature 174:647.

MacDowell EC, Allen E, MacDowell CG (1927). J Gen Physiol 11:57–70.

Markert CL, Petters RM (1978). Science 202:56–58.

Mayer JF, Fritz HI (1974). J Reprod Fertil 39:1–9.

McCance RA, Widdowson EM (1978). In "Human Growth." (F Falkner and JM Tanner, eds.), vol I, pp 145–166. Plenum, New York and London.

McLaren A (1965). J Reprod Fertil 9:79–98.

McLaren A (1976a). In "Embryogenesis in Mammals." (K Elliot and M O'Connor, eds.), Ciba Found Symp 40:47–51. Elsevier, Amsterdam.

McLaren A (1976b). "Mammalian Chimaeras." Cambridge University Press, Cambridge.

Minot CS (1891). J Physiol 12:7–31.

Moore, NW, Adams CE, Rowson LEA (1968). J Reprod Fertil 17:527–531.

Moore, RW, Eisen EJ, Ulberg LC (1970). Genetics 64:59–68.

Moustafa LA (1974). Proc Soc Exp Biol Med 147:485–488.

Mullen RJ, Whitten WK, Carter SC (1970). Annual Report of the Jackson Laboratory. Bar Harbor, Marine. pp 67–68.

Ounsted M, Ounsted C (1973). "On Fetal Growth Rate." Spastics International Medical Publications, London.

Parkinson CN (1958). "Parkinson's Law." Penguin Books, London.

Roberts RC (1966). Genet Res Camb 8:347–360.

Roberts RC, Falconer DS, Bowman P. Gauld IK (1976). Nature, 260:244–245.

Snow MHL (1973). Nature 244:513–515.

Snow MHL (1975). J Embryol Exp Morphol 34:707–721.

Snow MHL (1976a). J Embryol Exp Morphol 35:81–86.

Snow MHL (1976b). In "Embryogenesis in Mammals." (K Elliot and M O'Connor, eds.), Ciba Found Symp 40:53–70. Elsevier, Amsterdam.

Snow MHL (1977). J Embryol Exp Morphol 42:293–303.

Snow MHL (1978). In "Development in Mammals." (MH Johnson, ed.), vol III, pp 337–362. North-Holland, Amsterdam.

Snow MHL, Tam PPL (1979). Nature 279:555–557.

Tam PPL, Snow MHL (1981). J Embryol Exp Morphol (in press).

Tanner JM (1963). Nature, 199:845–850.

Tarkowski AK (1959a). Nature 184:1286–1287.

Tarkowski AK (1959b). Acta Theriol 3:191–267.

Tarkowski AK (1961). Nature 190:857–860.

Tarkowski AK, Witkowska A, Opas J (1977). J Embryol Exp Morphol 41:47–64.

Theiler K (1972). "The House Mouse." Springer-Verlag, Berlin, New York.

Tucker GD, Moor RM, Rowson LEA (1974). Immunology 26:613–621.

Walton A, Hammond J (1938). Proc R Soc B 125:311–335.

White JM, Legates JE, Eisen EJ (1968). Genetics 60:395–408.

Willadsen SM (1979). Nature 277:298–300.

Induction of the Heterogametic Gonad

Stephen S. Wachtel

Division of Cell Surface Immunogenetics, Memorial Sloan-Kettering Cancer Center
New York, New York 10021, and Division of Pediatric Endocrinology, New York
Hospital–Cornell Medical Center, New York, New York 10021

I. INTRODUCTION

H-Y antigen is a plasma membrane component identified by serum from female mice that have been sensitized with cells from males of the same highly

Levels of Genetic Control in Development, pages 219–234

220   **Wachtel**

TABLE I
Phylogenetic Conservatism of H-Y Antigen

Class	Species	Sex-chromosomes female/male	H-Y in
Mammals	Mouse	XX/XY	Male
	Man	XX/XY	Male
Birds	Chicken	ZW/ZZ	Female
Reptiles	Turtle	ZW/ZZ (?)[a]	Female
Amphibia	Leopard frog	XX/XY	Male
	Clawed frog	ZW/ZZ	Female
Osteichthyes	Medaka	XX/XY	Male[b]

[a]Female heterogamety inferred on the basis of H-Y phenotype
[Zaborski et al, 1979].
[b]See Pechan et al[1979].

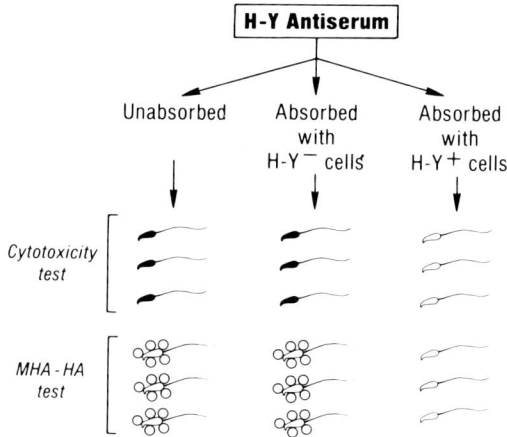

Fig. 1. H-Y typing by absorption. Positive absorption is manifested as a fall in percent sperm killed (stained) in the cytotoxicity test (top) and as a fall in percent labeled sperm (rosetted) in the MHA-HA test (bottom). The MHA-HA test uses a hybrid antibody reactive with mouse Ig on the one arm and with sheep red blood cells (SRBC) on the other arm. From Silvers and Wachtel [1977]. Copyright 1977 by the American Association for the Advancement of Science. Reprinted by permission.

inbred strain. It is called H-Y because the only source of incompatibility or antigenicity of male tissue in intrastrain females is the Y chromosome (which has no homolog in females), and because histocompatibility antigens in mice have generally been called H-1, H-2, H-3, H-X, and so on.

In 1971 Ellen Goldberg devised a serological assay for H-Y that has found routine applicability in our laboratory. According to Goldberg et al [1971], mouse epididymal sperm are incubated with H-Y antiserum (from male-grafted female mice) in the presence of rabbit complement. Sperm are killed as a consequence of their incubation with H-Y antiserum and complement, and dead cells are identified with trypan blue dye.

Specificity of the reaction is demonstrated by serological absorption. The H-Y antiserum is divided into equal portions: one portion is unabsorbed; one portion is absorbed with female spleen cells (for example); and one portion is absorbed with male spleen cells. (Absorbing cells are suspended in the antiserum for about a half-hour and then discarded.) Positive absorption, indicating presence of H-Y in the absorbing cells, is manifested as a fall in cytotoxic titer in the cytotoxicity test, and generally as a fall in reactivity in other similar systems (for example, the mixed hemadsorption-hybrid antibody test developed by Koo et al [1973]; Fig. 1).

Using the absorption technique, my colleagues and I detected presence of H-Y antigen or a highly cross-reactive molecule in a variety of species representing widely divergent pathways of vertebrate evolution. Because genes of fundamental significance rarely survive mutation, phylogenetic conservatism of H-Y antigen signaled conservation of a vital function. And because H-Y was detected in the heterogametic sex of each species (Table I), we proposed that H-Y antigen is the product of testis-determining genes in XY male species such as mouse and man, and that cross-reactive H-W antigen is the corresponding product of ovary-determining genes in ZW female species such as the chicken [Wachtel et al, 1975].

A corollary of that proposition is that testicular organogenesis in mammalian species could not occur in the absence of H-Y; presence of testicular architecture should signal corresponding presence of H-Y, the putative testis inducer, regardless of sex chromosome constitution or secondary sex phenotype. (That is because the sex-determining role of the mammalian Y-chromosome is limited to induction of the testis; secondary male differentiation is induced by testicular secretions [Jost, 1970]). Accordingly, we first set out to test the proposition by studying expression of H-Y in XX males, XX true hermaphrodites, and XY females. The following is a summary of data showing association of H-Y and testis and of H-W and ovary in vivo and in vitro and a discussion of the genetics and function of H-Y (H-W) antigens as indicated in studies conducted during the past several years.

II. H-Y AND THE TESTIS: IN VIVO CORRELATIONS

A. XX Males and XX True Hermaphrodites

Males lacking the Y-chromosome are known for several mammalian species. In the mouse, for example, the autosomal dominant gene *Sxr* causes differentiation of testes and subsequent male development in XX and XO embryos [Cattanach et al, 1971]; and in the goat, autosomal recessive genes in association with the *Polled* mutant give testes in XX embryos that subsequently exhibit a variety of phenotypes ranging from intersexual female to nearly "normal" male [Hamerton et al, 1969].

Similar conditions occur in man. In the family reported by Kasdan et al [1973] there was no evidence of a Y chromosome in either of two "brothers," one a 46,XX male with unambiguous testes, and the other a true hermaphrodite with scrotal ovotestes and a karyotype of 47,XX, including a marker chromosome—not a Y. A paternal uncle was also affected, exhibiting small aspermic testes and a karyotype of 47,XX with the marker. The pedigree indicated that sex reversal in this family was due to autosomal dominant genes. In another human family de la Chapelle et al [1978] discovered three XX males with a common ancestor born in 1664. Geneaological studies indicated that sex reversal in this case was due to autosomal recessive genes.

In another study, Evans and colleagues [1979] reported heteromorphic X chromosomes in nine of 13 46,XX males, inferring Y-to-X translocation of testis-determining genes in a corresponding majority of all human XX males. Indeed, XX sex reversal has been attributed to a variety of conditions, including Y-to-X translocation, abnormal activation of autosomal or X-linked testis-determining genes, Y-to-autosome translocation, XX/XY mosaicism, etc (see discussion in Wachtel and Bard [in press]). But whatever the indicated mode of inheritance, H-Y antigen has been detected in every case of XX sex reversal that we have studied (Table II).

A striking example of XX sex reversal is the XX true hermaphrodite, the individual possessing both ovarian and testicular tissue. Often these subjects have ambiguous genitalia. Accordingly, they have been recognized since antiquity, and it seems reasonable to assume that the *androgynos* described in the Talmud, for example, may well have satisfied our clinical definition of XX true hermaphroditism, in some cases at least. In the more modern case reported by Winters et al [1979], an ovotestis was discovered in the scrotum of a human true hermaphrodite with male habitus, female-like breasts, and scant facial hair. Cells from the ovarian and testicular portions of the gonad were cultured separately for cytogenetic and serologic analysis. The karyotype was 46,XX uniformly, but cells from the testicular portion were H-Y$^+$, and cells from the ovarian portion were H-Y$^-$.

TABLE II
H-Y Antigen in XX Males

Species	Mode of inheritance of testis-determining genes	H-Y phenotype	Reference
Mouse	Autosomal dominant (Sxr)	+	Bennett et al [1977]
Goat	Autosomal recessive (P)	+	Wachtel et al [1978]
Dog	Autosomal[a]	+	Selden et al [1978]
Human	Autosomal dominant	NT	Kasdan et al [1973]
	Autosomal recessive	+	de la Chapelle et al [1978]
	Y-X interchange	+	Breg et al [1979]

[a]It remains to be ascertained whether this is a dominant or a recessive mutation
NT = not tested

B. XY Females

The wood lemming, Myopus schisticolor, is remarkable for a 4:1 sex ratio favoring the female. About half the females have a normal male karyotype (32,XY), yet they are fertile. In that connection two points are worth mentioning: First, XY females of the wood lemming are H-Y$^-$; and second, the XY female condition is inherited as an X-linked trait. In fact two kinds of X chromosomes in the wood lemming are distinguished by G-banding. One is found in XY females and the other in XY males [Herbst et al, 1978].

Whereas a single X chromosome is sufficient for development of the fertile ovary in rodents, two X chromosomes are required to sustain the differentiated ovary in humans. In the absence of the second X chromosome, as in 45,X Turner syndrome, the human gonad starts out as an ovary, but the germ cells die and the ovary degenerates. According to that scheme, XY human gonads lacking a functional testis-determinant should initially organize ovaries, but the ovaries should degenerate around the time of birth or shortly thereafter. The clinical condition is called XY gonadal dysgenesis. Consider the following study: Dysgenetic ovaries were discovered in a grossly retarded phenotypic female with the karyotype 46,Xp$^+$Y (signifying additional material in the short arm, p, of the X chromosome), and normal fetal ovaries were later discovered in her 46,Xp$^+$Y female sibling (an abortus of 20 weeks' gestation). The propositus and her fetal sib were typed H-Y$^-$ [Bernstein et al, 1980].

In a related clinical study, Rosenfeld et al [1979] discovered dysgenetic ovaries in a phenotypic female with some of the stigmata of Turner syndrome and the karyotype 46,XYp$^-$, signifying deletion of the Y short arm in this case. Cells from Rosenfeld's patient were also typed H-Y$^-$ in comparison with cells from normal XY males. So the two cases taken together provide evidence not only for a testis-inducing role of H-Y antigen but also for two genes in testicular differentiation, one on the X chromosome and the other on the Y. (For discussion of H-Y genetics, see section IV.)

It is remarkable that of some 30 cases of 46,XY gonadal dysgenesis in human phenotypic females studied for H-Y, two-thirds were typed H-Y$^+$ [see, for example, Wolf, 1979]. At first sight, presence of H-Y in some cases of XY gonadal dysgenesis and absence of H-Y in others might seem paradoxical, but the paradox is resolved if one considers the likelihood that H-Y is *functionally* absent in both conditions. There is considerable evidence for the existence of a specific H-Y antigen receptor. If testicular differentiation of the indifferent embryonic gonad were secondary to reaction of H-Y and its receptor, then ovarian differentiation should ensue in cases of receptor failure, despite presence of H-Y in the plasma membrane of somatc cells (see section VB).

III. H-Y AND THE TESTIS: IN VITRO CORRELATIONS

A. XY Testicular Cells Lysostripped of H-Y Antigen Form Ovarian Aggregates

To a limited degree, XX and XY sex reversal have now been accomplished in the laboratory. In the first direct demonstrations of a testis-organizing function of H-Y antigen, dispersed XY cells from the testes of normal newborn mice or rats were cultured in the presence or absence of specific H-Y antibody [Ohno et al, 1978; Zenzes et al, 1978c]. The rationale was as follows: 1) When cells are cultured under conditions of very slow rotation, they reaggregate to form structures characteristic of the tissue from which they have been taken [Moscona, 1961]. Under these conditions dispersed cells of the testicular tubule might be expected to organize tubular aggregates, for instance. 2) If a particular cell particular cell surface molecule were critical to formation of the tubule, then formation of the tubule could not occur in the absence of that cell surface molecule. 3) In the presence of specific antibody, cell surface antigens migrate to a polar cap of the cell where they are internalized and digested by autophagic lysosomes (hence they are said to be "lysostripped" in the process). 4) Thus if H-Y antigen were critical to organization of the seminiferous tubule, testicular reaggregation could not occur in suspensions of XY testicular cells following their exposure to excess H-Y antibody.

The rationale is vindicated in practice. Dispersed XY testicular cells of mouse or rat organize long tubular structures in overnight rotary cultures. But when they are exposed to specific H-Y antibody, the same cells organize spherical aggregates resembling the follicles of the newborn ovary.

B. XX Ovarian Cells Organize Seminiferous Tubules on Exposure to H-Y Antigen

In reciprocal experiments, Zenzes et al [1978b] cultured XX cells from the dissociated ovaries of newborn rats in medium from cultivated testicular cells, a putative source of soluble H-Y antigen. In this case the pattern of reaggregation assumed tubular proportions. And in a related study, Muller et al [1978b] found hCG receptors in the XX converted gonadal cells. The hCG receptors are normally found in rat testicular cells at birth, but they do not appear in cells of the rat ovary until six to eight days after birth.

Rather a striking demonstration of in vitro "sex reversal" was obtained in the whole organ culture experiments of Ohno et al [1979]. Indifferent XX gonads of the fetal cow of 27 mm crown–rump length were cultured in a concentrated source of H-Y (from cultured Daudi cells; see section VA). After five days the gonads were transformed into small testes containing well-defined seminiferous tubules and tunica albuginea.

IV. GENETICS OF H-Y ANTIGEN

A. Evidence for Y-Linked Genes

Increased expression of H-Y antigen in the white blood cells of XXYY and XYY males in comparison with its expression in normal XY males signaled occurrence of a gene for H-Y on the Y chromosome. The gene was mapped by correlating expression or non-expression of H-Y with presence or absence of particular portions of the Y in patients exhibiting structural or numerical abnormalities of the Y. The gene was located on the short arm near the centromere; in a single case, a pericentric locus on the long arm could not be ruled out [Koo et al, 1977]. The Y-chromosomal H-Y gene loci were the same as testis-determining gene loci identified in an earlier study [Simpson, 1975].

B. Evidence for X-Linked Genes

We have already pointed out that mutation of the X chromosome can be associated with non-expression of H-Y antigen, with failure of testicular differentiation, and thereby with XY sex reversal in rodents and in man. Evidence of a role of the X chromosome in testicular differentiation is also provided in females with Turner syndrome and the karyotypes 46,X,i(Xq) or 45,X/46,X,i(Xq), symbolizing presence of at least a single cell line with an isochromosome (duplication) of the long arm of the X [hence:i(Xq)]. H-Y antigen has been detected in such cases in reduced amounts, for example by Wolf et al [1980]. The inference is that H-Y genes are induced during the structural rearrangements leading to formation of the isochromosome and that these genes are situated on the non-inactivated portion of the X.

C. Evidence for Autosomal Genes

The testis-determining autosomal mutants described in section IIA (*Sxr* of the mouse, *Polled* of the goat, and their counterparts in man) provide ample evidence of testis-determining H-Y genes that are neither Y-situated nor X-linked. For the present, however, it is not clear whether these genes are translocated from the Y (or X), whether they are mutants that mimic the Y (or X), or whether they are constitutive mutants of *pre-existing* testis-determinants.

There is reason to believe that *Sxr* may have arisen as a Y-to-autosome translocation. Laura Tres [personal communication] recently found association of the pericentric Yp and an unidentified autosomal segment (indicating common nucleotide base sequence) in pachytene spermatocytes from *Sxr/ −* , XY carrier males (the fathers of *Sxr/ −* , XX sex-reversed males). Still this does not tell us whether the *Sxr* mutant is a structural or regulatory H-Y gene; one might argue that the structural testis-determining H-Y gene is on *another* autosome.

D. Where Is the H-Y Structural Gene?

The locus of the structural H-Y gene remains to be ascertained. If the structural locus were situated on the Y, then XX sex revesal could be due to Y-autosome or Y-X translocation, but it is not clear how a translocated segment of the Y could act as a dominant gene in the mouse and as a recessive gene in the goat. (An explanation is that there is a *series* of H-Y genes on the Y, and that dominant versus recessive modes of sex reversal are determined by the portion of genes translocated in a given mutational event [Wachtel et al, 1978].) Moreover, it is not clear why a Y-X translocation should masquerade as an isochromosome of Xq or why this should involve production of a defunct H-Y, unable to support testicular differentiation [Wachtel et al, 1980c].

If the structural H-Y gene were X-linked [Hamerton et al, 1969], then XX sex reversal could be due to constitutive synthesis of H-Y antigen in the absence of a putative Y activator. Indeed, XX male sex reversal and XX true hermaphroditism might represent different aspects of X chromosome inactivation ("lyonization"), ovarian versus testicular differentiation depending on the frequency of gonadal cells carrying the noninactivated X mutant. In that connection it is worth noting that XX human true hermaphrodites have H-Y intermediate blood types compared with the H-Y+ blood types of normal XY males, and that

XX males have blood types that may vary from H-Y$^+$ to H-Y intermediate. But the question of dominant versus autosomal modes of XX sex reversal is not easily resolved in this model; and it is still not clear why a mutant i(Xq) should cause production of a "defunct" species unable to support testicular differentiation.

What if the structural H-Y gene were autosomal? According to the model proposed by Wolf et al [1980], the autosomal H-Y gene pair is repressed by the X chromosome pair (the repressor genes being situated on the noninactivated region of the X short arm) and de-repressed by genes on the Y chromosome. It would follow that dominant versus recessive modes of XX sex reversal merely represent different degrees of escape from X-mediated suppression. Moreover, H-Y synthesis would be expected in cases of 46,X,i(Xq), and indeed in almost any female lacking the normal X pair. In support of that scheme Wolf et al [1980] report "intermediate" expression of H-Y in females with 45,X Turner syndrome, but the issue is complicated by Haseltine [personal communication], who fails to detect H-Y in the same condition (45,X Turner syndrome) using the same assay.

V.   CELL-BOUND H-Y AND SOLUBLE H-Y

A. Membrane-Associated H-Y, the Transplantation Antigen

H-Y antigen was discovered about 25 years ago with the observation that male-to-female skin grafts are unsuccessful in certain highly inbred strains of the mouse [Eichwald and Silmser, 1955]. Rejection times were slow: about 25 days compared with rejection times of eight to 12 days for skin grafts exchanged between mice differing at the "major" H-2 locus. Male-to-female incompatibility thus seemed to represent a single weak immunogenetic difference between donor and host in an otherwise isogenic system. Quite naturally a number of studies were undertaken to clarify the genetics of male-to-female incompatibility. Among the questions raised were the following: Is H-Y transplantation antigen a Y-chromosome-determined or a hormone-mediated trait? Why do some females accept male grafts? Is H-Y a characteristic of *all* male mice?

One of the points to emerge from these studies was that expression of H-Y is modified by other cell surface antigens, namely those determined by the major histocompatibility complex (MHC), the H-2 locus of mice. In vivo, for example, H-2^k male grafts are rejected faster than H-2^b male grafts by H-2^k/H-2^b females [Silvers and Billingham, 1967] (see Table III); and in vitro, cell-mediated cytotoxic reactions against H-Y incompatible target cells are H-2 restricted, such that killing depends in part on H-2 haplotype of the male target cells [Gordon and Simpson, 1976].

TABLE III
**Effect of Donor H-2 Haplotype on Survival of H-Y Incompatible Skin Grafts in (B10 ×
B10·BR)F₁ Female Hosts***

Group	Male donor	Number	Donor H-2	Recipient H-2	MST[a] ± SD
A	B10	30	b	b/k	20.0 ± 1.4
B	B10·BR	30	k	b/k	14.5 ± 1.3
C	(B10 × B10·BR)F₁	14	b/k	b/k	17.5 ± 1.3

*The source of histoincompatibility in these donor-host combinations is the Y chromosome, which is the same in all three groups. Since strains B10 and B10·BR are congenic pairs varying *only* with respect to H-2 haplotype, the difference in graft survival must be due to differential presentation of donor H-2 in association with H-Y. Thus the data indicate physical association of H-2 and H-Y antigens on the cell surface [Wachtel et al, 1973].
[a]MST = median survival time.

These observations pointed to a physical association of H-2 and H-Y; on that basis Ohno [1977] suggested that H-2 antigen (heavy chains) and beta-2-microglobulin (light chains) serve conjointly as *anchorage sites* for H-Y antigens, and indeed for all organogenesis-directing proteins in all tissues. According to that scheme, H-Y is not itself integrated into the membrane and requires a stable membrane anchorage site or "carrier." A corollary is that H-Y could not be expressed on the somatic cell surface in the absence of its β_2m-MHC carrier and would be secreted instead into the extracellular milieu. So it is remarkable that cultured "Daudi" cells, which are β_2m($-$) and MHC($-$), in fact secrete H-Y into the medium [Beutler et al, 1978].

Readers should note that there is another cultured line, "Chevalier," β_2m($+$) and MHC($-$), that retains membrane-bound H-Y [Fellous et al, 1978]; that there is a variant Daudi line, β_2m($-$) and MHC($-$) that also retains membrane-bound H-Y [Iwata et al, 1979]; and that H-Y and MHC antigens have been mapped separately on the surface of the mouse thymocyte [Flaherty et al, 1979]. So the notion that β_2m-MHC dimers bind H-Y to the membrane remains to be clarified. Nevertheless, the notion has important implications for H-Y typing. Von Boehmer et al [1979] have reported that in mice killer T-cells can mistake H-2D^d antigens for (H-Y + H-2D^b). This raises the specter of false-positive or false-negative results in clinical situations, requiring careful evaluation of the diagnostic use of H-Y serology.

B. The Specific H-Y Receptor

H-Y antigen is ubiquitous, being found in all tissues that have been tested in the normal male. The putative β_2m-MHC carrier is also ubiquitous. It follows that H-Y–mediated testicular organogenesis requires a specific H-Y receptor limited to the gonadal primordium. Existence of a specific H-Y receptor is implicit in several studies, notably those involving complete testicular transformation in XX/XY chimeric gonads in vivo, and of XX gonads in vivo and in vitro. And, indeed, the receptor has been demonstrated by reacting soluble H-Y from Daudi cells or from rodent testicular cells with cells from gonadal and extragonadal tissues. The rationale of these experiments was as follows: 1) Female cells possessing a receptor for H-Y should acquire H-Y if it is present. 2) These cells should bind specific H-Y antibody in serological tests and so should type H-Y$^+$. Thus Muller et al [1978a] observed specific uptake of H-Y antigen in ovary cells (and to a lesser extent in testicular cells presumably saturated with indigenous H-Y) but not in the extragonadal tissues.

C. A Hormone-Like Role of H-Y in the Bovine Freemartin

In cattle and the other Bovidae, binovular twinning is notable for the establishment of chorionic vascular anastomoses between fetuses. In cases of heterosexual twinning, this is associated with masculinization of the female twin or *freemartin,* and in extreme situations, with testicular transformation of the freemartin gonad (Fig. 2). In fact, masculinization of the freemartin may be due in large part to androgens secreted in her own gonad [Short et al, 1969]. In 1976 Ohno et al discovered that H-Y antigen is prominent in the gonad of the freemartin. It was proposed that H-Y is disseminated in "hormone-like" fashion by migratory XY cells and taken up by XX gonadal cells, which are thereby induced for testicular differentiation. The proposal was challenged, however, by the report that freemartinism did not occur when vascular connections between male and female were severed between days 37 and 45 of gestation despite presence of XY cells in liver, and by inference, in the other tissues of the female [Vigier et al, 1976].

On the basis of more recent work, we now believe that H-Y is disseminated in the bull twin, transported in the serum, and bound by gonadal receptors in the female. Thus bovine fetal ovarian cells acquire the H-Y$^+$ cellular phenotype on exposure to fetal calf serum from bull or freemartin but not normal cow; and in the new "competitive binding radio-assay," specific reaction of tritiated H-Y and its fetal ovarian receptor is inhibited by fetal calf serum from bull or freemartin, but not normal cow [Wachtel et al, 1980b].

D. H-Y Is Secreted by Testicular Sertoli Cells

H-Y is actively secreted by Sertoli cells [Zenzes et al, 1978a]. Using specific H-Y antibody and goat anti-mouse IgG, John Hall of our laboratory has now immunoprecipitated radioactive H-Y from the supernatant fluid of Sertoli cell

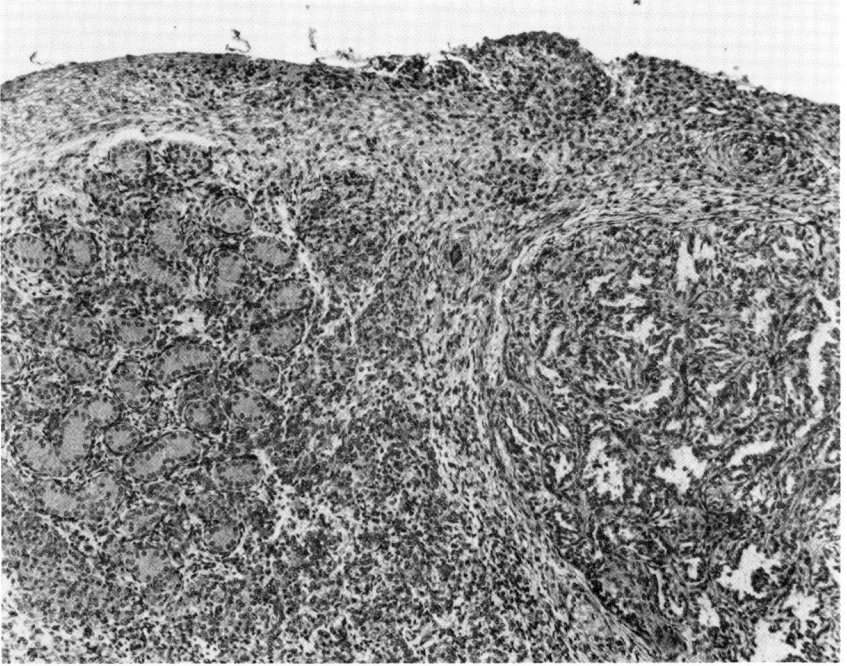

Fig. 2. Testicular differentiation in freemartin gonad. Highly organized testicular tubules, rete testis, and tunica albuginea in gonad of H-Y$^+$ bovine fetal freemartin of about 160 days gestation (35 cm crown–rump length; magnification, × 40). From Wachtel et al [1980c]. Copyright 1980 by the MIT Press. Reprinted by permission.

cultures. A profile of secreted proteins was established by running the precipitated proteins on a polyacrylamide gel, slicing the gel, and individually solubilizing the slices under reducing conditions (SDS). Three peaks were obtained: one representing high molecular weight material that was not resolved in this system; one representing a polypeptide of molecular weight 31,000; and one, a polypeptide of molecular weight 16,500, the last corresponding to Daudi-secreted H-Y (mol wt ≃ 16,000–18,000). It should be mentioned that radioiodination and solubilization of Sertoli cell surface proteins likewise yields two peaks representing polypeptides with molecular weights of 31,000 and 16,000–18,000 [Hall and Wachtel, in press].

VI. H-Y (H-W) ANTIGEN AND THE HETEROGAMETIC OVARY

In females of the chicken only the left gonad develops as an ovary. The right gonad develops as an ovary initially, but within a day or two of hatching, the right ovary degenerates. If the left ovary is destroyed experimentally or by disease, the right residual gonad may become a testis, and the chicken may become an impotent rooster, hence the old English saying, "a whistling woman and a crowing hen are neither good for gods nor men."

There is evidence that mammalian H-Y antigen may be ovary-determining in birds, and indeed in other ZW/ZZ vertebrate species as well: 1) H-Y (H-W) is present in the left gonad of the chick hatchling, but not to the same extent in the right gonad (our preliminary observations); 2) XY mammalian testes co-cultured with avian ZZ testes induce ovarian differentiation in the latter [Akram and Weniger, 1968]*; 3) estradiol-mediated sex reversal of ZZ embryos of the chicken and ZZ tadpoles of the clawed frog, Xenopus laevis, is associated with acquisition of the H-W+ *gonadal* phenotype in both species [Muller et al, 1979; Wachtel et al, 1980a]; and finally, 4) although it might be argued that cross-reactive H-Y and H-W antigens are only representative parts of larger, functionally disparate molecules, our studies of the clawed frog and the domestic chicken have revealed conservatism of the gonadal H-Y (H-W) receptor. Soluble H-Y of the mouse testis is readily taken up by ZZ testicular cells in both these female heterogametic species, so we infer that the active site of the H-Y molecule is the same as the active site of the H-W molecule.

How then does the same molecule induce the XY testis in one species and the ZW ovary in another? Consider the following: 1) H-Y must be directly involved in maintenance of the seminiferous tubule, because XY Sertoli cells fail to organize tubules when they have been lysostripped of H-Y in vitro. 2) An essential difference between the left ovary and the right gonad in ZW hatchling chicks is presence of germ cells in the former and a relative paucity of germ cells in the latter. This is correlated with occurrence of medullary cords in the hatchling right gonad that bear a striking resemblance to the seminiferous tubules of the normal testis (due to residual "ZZ testis-inducer" or, possibly, to residual H-W). 3) Of all nucleated male cells in the mammal, only premeiotic germ cells are H-Y−. Mammalian male germ cells do not enter meiosis until puberty; it is at that time that they assume the H-Y+ phenotype [Zenzes et al, 1978a; Koo et al, 1979].

If H-Y is physically involved in maintenance of gonadal architecture, and if mammalian premeiotic germ cells lack H-Y (and its receptor [Zenzes et al, 1978a]), then they cannot contribute to initial development of the testis, to wit, organization of seminiferous tubules and tunica albuginea. Since germ cells of the ZW female chick are engaged in meiosis before hatching, it follows that the hatchling oocyte may be H-W+ and perhaps receptor positive. Thus if H-Y is physically involved in maintenance of the ZW ovary, the female germ cell *can*

---

*This could be due to activation of H-W genes by estrogens synthesized in the presence of "excess" male steroids [Muller et al, 1979; Ohno and Matsunaga, 1981].

contribute to development of the ovary, to wit, organization of the follicle. Indeed the common (H-Y$^+$/receptor$^+$) granulosa-Sertoli cell precursors [Ciccarese and Ohno, 1978] might be expected to form links with the central germ cell as well as with each other, thereby generating a spherical aggregate rather than the cylindrical or tubular aggregate attributed by Ohno [1979] to bipolar distribution of mammalian H-Y and consequent end-to-end linkage of Sertoli cells [and see Ohno and Matsunaga, 1981].

The model requires a yet-to-be discovered ZZ testis-inducer in the non-mammalian vertebrates, and a corresponding XX ovary-inducer in the mammalian species. In that connection, it is remarkable that neonatal thymectomy causes ovarian degeneration in mice [Nishizuka and Sakakura, 1971], and that that is associated with production of circulating autoantibody directed against oocyte antigens [Taguchi et al, in press].

VII. CONCLUSIONS

In mammals differentiation of the testis may be attributed to cell surface expression of H-Y antigen, and aberrant differentiation of the testis may be ascribed to absence of H-Y or its gonadal receptor. For example H-Y causes differentiation of testicular tubules in XX indifferent gonads in organ culture, and H-Y antibody blocks testicular reaggregation of dispersed XY Sertoli cells in rotary culture.

In birds and some amphibians differentiation of the ovary may be attributed to cell surface expression of serologically cross-reactive H-W antigen. For example H-W is present in the functional left ovary of the newly hatched chick, but its expression is reduced in the degenerative right gonad; and estradiol-induced sex reversal is correlated with appearance of H-W in sex-reversed ZZ gonads of the chicken and clawed frog.

Based on these considerations, a common differentiative mechanism may be inferred in organogenesis of the vertebrate heterogametic gonad. According to one scheme, H-Y synthesized in the presumptive XY testis acts via the specific receptor to link precursors of the common Sertoli–granulosa cell lineage, thereby generating end-to-end association of developing Sertoli cells in the seminiferous tubule; germ cells, which have not yet entered meiosis, do not partake in this differentiative event. In the corresponding stages of avian embryogenesis, H-W synthesized in the presumptive ZW ovary acts in a similar fashion to link precursors of the Sertoli-granulosa cell lineage, but in this case the germ cells enter meiosis much earlier and thus synthesize H-W and/or its receptor. Avian germ cells may therefore actively participate in follicular morphogenesis; presumably the result is spherical aggregation of developing granulosa (follicular) cells around the central H-W$^+$/receptor$^+$ oocytes.

ACKNOWLEDGMENTS

This work was supported in part by grants from the N.I.H. (AI 11982, CA 08748, HD 10065, and HD 00171) and by grant 6-247 from The March of Dimes Birth Defects Foundation.

REFERENCES

Akram H, Weniger JP (1968). Arch Anat Microsc Morphol Exp 57:369–378.

Bennett D, Mathieson BJ, Scheid M, Yanagisawa K, Boyse EA, Wachtel SS, Cattanach BM (1977). Nature 265:255–257.

Bernstein R, Koo GC, Wachtel SS (1980). Science 207:768–769.

Beutler B, Nagai Y, Ohno S, Klein G, Shapiro IM (1978). Cell 13:509–513.

von Boehmer H, Hengartner H, Nabholz M, Lernhardt W, Schreier MH, Haas W (1979). Eur J Immunol 9:592–597.

Breg WR, Genel M, Koo GC, Wachtel SS, Krupen-Brown K, Miller OJ (1979). In "Genetic Mechanisms of Sexual Development." (HL Vallet and IH Porter, eds.), pp. 279–292. New York.

Cattanach BM, Pollard CE, Hawkes SG (1971). Cytogenetics 10:318–337.

de la Chapelle A, Koo GC, Wachtel SS (1978). Cell 15:837–842.

Ciccarese S, Ohno S (1978). Cell 13:643–650.

Eichwald EJ, Silmser CR (1955). Transplant Bull 2:148–149.

Evans HJ, Buckton KE, Spowart G, Carothers AD (1979). Hum Genet 49:11–31.

Fellous M, Gunther E, Kemler R, Wiels J, Berger R, Guenet JL, Jakob H, Jacob F (1978). J Exp Med 147:58–70.

Flaherty L, Zimmerman D, Wachtel SS (1979). J Exp Med 150:1020–1027.

Goldberg EH, Boyse EA, Bennett D, Scheid M, Carswell EA (1971). Nature 232:478–480.

Gordon RD, Simpson E (1976). In "Proceedings of the Tenth Leucocyte Culture Conference." (VP Eijsvoogel, D Roos, WP Zeylemaker, eds.), p 521. Academic Press, New York.

Hall JL, Wachtel SS (in press). Mol Cell Biochem.

Hamerton JL, Dickson JM, Pollard CE, Grieves SA, Short RV (1969). J Reprod Fertil (Suppl 7):25–51.

Herbst EW, Fredga K, Frank F, Winking H, Gropp A (1978). Chromosoma 69:185–191.

Iwata H, Nagai Y, Stapleton DD, Smith RC, Ohno S (1979). Arthritis Rheum 22:1211–1216.

Jost A (1970). Phil Trans R Soc Lond B 259:119–130.

Kasdan R, Nankin HR, Troen P, Wald N, Pan S, Yanaihara T (1973). N Engl J Med 288:539–545.

Koo GC, Stackpole CW, Boyse EA, Hammerling U, Lardis M (1973). Proc Natl Acad Sci USA 70:1502–1505.

Koo GC, Wachtel SS, Krupen-Brown K, Mittl LR, Breg WR, Genel M, Rosenthal IM, Borgaonkar DS, Miller DA, Tantravahi R, Schreck RR, Erlanger BF, Miller OJ (1977). Science 198:940–942.

Koo GC, Mittl LR, Goldberg CL (1979). Immunogenetics 9:293–296.

Moscona A (1961). Exp Cell Res 22:455–475.

Muller U, Aschmoneit I, Zenzes MT, Wolf U (1978a). Hum Genet 43:151–157.

Muller U, Zenzes MT, Bauknecht T, Wolf U, Siebers JW, Engel W (1978b). Hum Genet 45:203–207.

Muller U, Zenzes MT, Wolf U, Engel W, Weniger J-P (1979). Nature 280:142–144.

Nishizuka Y, Sakakura T (1971). Endocrinology 89:886–893.

Ohno S (1977). Immunol Rev 33:59–69.

Ohno S (1979). "Major Sex-Determining Genes." Springer Verlag, New York.

Ohno S, Christian LC, Wachtel SS, Koo GC (1976). Nature 261:597–599.

Ohno S, Nagai Y, Ciccarese S (1978). Cytogenet Cell Genet 20:351–364.

Ohno S, Nagai Y, Ciccarese S, Iwata H (1979). Recent Prog Horm Res 35:449–470.

Ohno S, Matsunaga T (1981). In "The 39th Symposium of the Society for Developmental Biology." (S. Subtelny and UK Abbott, eds.), pp 235–246. Alan R. Liss, New York.

Pechan P, Wachtel SS, Reinboth R (1979): Differentiation 14:189–192.

Rosenfeld RG, Luzzatti L, Hintz RL, Miller OJ, Koo GC, Wachtel SS (1979): Am J Hum Genet 31:458–468.

Selden JR, Wachtel SS, Koo GC, Haskins ME, Patterson DF (1978): Science 201:644–646.

Short RV, Smith J, Mann T, Evans EP, Hallett J, Fryer A, Hamerton JL (1969): Cytogenetics 8:369–388.

Silvers WK, Billingham RE (1967): Science 158:118–119.

Silvers WK, Wachtel SS (1977): Science 195:956–960.

Simpson JL (1975): Birth Defects: Orig. Art. Ser. 11:23–59.

Taguchi O, Nishizuka Y, Sakakura T, Kojima A (in press): Clin Exp Immunol.

Vigier B, Locatelli A, Prepin J, du Mesnil du Buisson F, Jost A (1976): CR Acad Sci (Paris) 282:1355–1358.

Wachtel SS, Bard J (in press). In "The Intersex Child". (N. Josso, ed.) S. Karger, Basel.

Wachtel SS, Gasser DL, Silvers WK (1973). Science 181:862–863.

Wachtel SS, Ohno S, Koo GC, Boyse EA (1975). Nature 257:235–236.

Wachtel SS, Basrur P, Koo GC (1978). Cell 15:279–281.

Wachtel SS, Bresler PA, Koide SS (1980 a). Cell 20:859–864.

Wachtel SS, Hall JL, Muller U, Chaganti RSK (1980 b). Cell 21:917–926.

Wachtel SS, Koo GC, Breg WR, Gene M (1980 c). Hum Genet 56:183–187.

Winters SJ, Wachtel SS, White BJ, Koo GC, Javadpour N, Loriaux L, Sherins RJ (1979). N Engl J Med 300:745–749.

Wolf U (1979). Hum Genet 47:269–277.

Wolf U, Fraccaro M, Mayerova A, Hecht T, Zuffardi O, Hameister H (1980). Hum Genet 54:315–318.

Zaborski P, Dorizzi M, Pieau C (1979). CR Acad Sci (D) (Paris) 288:351–354.

Zenzes MT, Muller U, Aschmoneit I, Wolf U (1978a). Hum Genet 45:297–303.

Zenzes MT, Wolf U, Engel W (1978b). Hum Genet 44:333–338.

Zenzes MT, Wolf U, Gunther E, Engel W (1978c). Cytogenet Cell Genet 20:365–372.

The Role of H-Y Plasma Membrane Antigen in the Evolution of the Chromosomal Sex Determining Mechanism

Susumu Ohno and Takeshi Matsunaga

Division of Biology, City of Hope Research Institute, Duarte, California 91010

I. INTRODUCTION

The conspicuously heteromorphic XY sex chromosome pair that characterizes the heterogametic male sex of all mammals does not symbolize the modern and most advanced form of sexual reproduction. On the contrary, sex chromosomes are very ancient in their origin. In the plant kingdom, the large X and the small Y chromosome, as heteromorphic as their mammalian counterparts are commonly found among mosses such as Sphaerocarpus donnelli [Allen, 1917] of the primitive phylum, Bryophyta. The true hermaphroditism of modern Spermatophyta embodied in the one flower bearing pollen on the stamen and ovules in pistils may have developed after a later abandonment of the chromosomal sex-determining mechanism by higher plants. Among vertebrates, the conspicuously heteromorphic sex chromosomes are already found in certain teleost species [Ebeling and Chen, 1970], although a greater number is endowed with the homomorphic sex chromosome pair, the X and the Y or the Z and the W remaining morphologically as well as genetically almost identical.

Levels of Genetic Control in Development, pages 235–246

Actually, both the heteromorphic and homomorphic forms of sex chromosome pairs serve the same end: to keep a genetic difference between the two sexes at a necessary minimum, the former by letting the Y or the W chromosome undergo extensive genic degeneration until it retains but a few sex-determining genes, and the latter by keeping the X and the Y or the Z and the W essentially as a pair of homologues, in spite of having opposite sex-determining alleles at a few critical gene loci. As the specialized, but degenerate Y or the similar W can no longer substitute for the X or the Z, the former form of sex chromosomes imposed status quo on the existing sex-determining mechanism. The X-inactivation mechanism operating in mammalian females can be viewed as a secondary attempt to nullify a genetic disparity between the two sexes within the imposed status quo [Lyon, 1961]. By contrast, those with the homomorphic form enjoy further options to change their sex-determining mechanism via WW or YY individuals, which are quite viable and fertile [Yamamoto, 1961].

Indeed, teleost fish species endowed with the latter form are capable of switching their chromosomal sex-determining mechanism from XX/XY male heterogamety to the ZW/ZZ female heterogamety and vice versa [Gordon, 1951]. Hermaphroditic species of both the synchronous and asynchronous types found in divergent genera of teleost fish [Chan, 1970] likely represent other options exercised by gonochorist species with the latter form of sex chromosomes.

In the normal course of embryonic development, the fate of gonads determines the sex of an individual, and Moscona's classic experiment revealed organogenesis to be the consequence of specific plasma membrane interactions among component cells [Moscona, 1957]. The antiquity of the chromosomal sex-determining mechanism indicates that the same set of plasma membrane components has been organizing vertebrate gonads from time immemorial. The two pairs of plasma membrane components constitute the minimal requirement for the testicular versus ovarian alternative of gonadal organogenesis; they are a testis-organizing agent and its specific receptor, as well as an ovary-organizing agent and its specific receptor. In all vertebrate classes, excluding mammals, however, both genetic sexes apparently are capable of expressing all four plasma membrane components for gonadal organogenesis. Accordingly, complete sex reversals to either direction can easily be induced in fish [Yamamoto, 1961] and amphibians [Mikamo and Witschi, 1964] by environmental manipulations that include steroid hormone treatments. Among reptiles, the sex of individual turtles is apparently determined by the incubation temperature of hatching eggs [Pieau, 1975]. Although birds are already equipped with the conspicuously heteromorphic ZW sex chromosome pair, sex reversals can still be achieved. In chickens or pheasants, an early removal of the functional left ovary from ZW genetic females causes the right residual gonad to transform to a testis [Miller, 1938], and injections of estradiol-17β or its precursor to hatching eggs cause ZZ testes to tranform to ovotestes [Wolff and Ginglinger, 1935]. Only in placental mammals has the expression of one of the four gonadal organogenesis components become inflexibly confined to the genetic male. An evolutionary reason for this inflexibility in expression is readily found in the need of mammalian male embryos to differentiate in the entirely feminine environment of the mother's womb. The proposal that identified this male specific agent of mammalian gonadal organogenesis as H-Y plasma membrane antigen [Wachtel et al, 1978] has since been supported by a large body of experimental evidence. The precise role this antigen plays in mammalian gonadal organogenesis, however, can be better understood if we consider its history during the long course of vertebrate evolution.

II. REGULATORY ECONOMY IN GONADAL ORGANOGENESIS

An embryonic gonadal primordium of mammals develops on the mesonephric surface facing the dorsal mesentery. Somatic elements of the gonad are thought to be derived from either mesonephric duct epithelia or parietal epithelia of mesonephric glomeruli. Primordial germ cells, by contrast, are migrants from the yolk sac.

The fact that both genetic sexes of earlier vertebrates were potentially capable of organizing either testes or ovaries enables mammals to exercise a considerable economy in regulation of gonadal organogenesis. As the expression of testis-organizing H-Y antigen became inflexibly male specific and beyond the reach of either inductive or suppressive agents in mammals, it began to be expressed by early male zygotic nuclei, together with a host of proteins for cellular household chores—ie, at the eight-cell stage of male mouse embryos [Krco and Goldberg, 1976]. Inevitably, its subsequent expression became ubiquitous as opposed to cell-type specific or organ specific. The expression of plasma membrane receptor sites for H-Y antigen, however, is confined to gonadal cells as it must always have been, but of both sexes [Ohno, 1976; Muller et al, 1979a; Nagai et al, 1979]. Thus, we see the first example of regulatory economy in mammalian gonadal organogenesis that utilized the past history of individual components. Since the expression of its specific receptor is confined to gonadal cells, that of H-Y antigen itself can afford to be ubiquitous. Inasmuch as the expression of H-Y antigen is constitutively confined to genetic males, it is a regulatory waste to confine the expression of gonad-specific receptor for H-Y antigen also to genetic males. In Moscona-type reaggregation experiments, a free suspension of newborn rodent testicular cells lysostripped of their H-Y antigen readily reorganized ovarian follicle-like structures [Ohno et al, 1978a; Zenzes et al, 1978a]. The above apparently means that the mammalian genetic males retained the inherent capability of their predecessors to express an ovary-organizing antigen as well as its gonad-specific receptor. It is H-Y antigen's

TABLE I
Expression of Four Plasma Membrane Components for Mammalian Gonadal Organogenesis

	XY male			XX female		
	Germ	Gonadal soma	Extragonadal soma	Germ	Gonadal soma	Extragonadal soma
Testis-organizing H-Y antigen	(−) Induced at puberty	(+)	(+)	(−)	(−)	(−)
Its specific receptor	(−) Induced at puberty	(+)	(−)	(−)	(+)	(−)
Ovary-organizing ? antigen	(+)?	(+)? Suppressible by H-Y	(+)?	(+)?	(+)?	(+)?
Its specific receptor	(+)?	(+)? Suppressible by H-Y	(−)?	(+)?	(+)?	(−)?

The distribution of testis-organizing H-Y antigen and its specific receptor is factual, whereas that of an ovary-organizing antigen and its specific receptor is inferred. The very fact that a mere removal of H-Y antigen causes neonatal testicular cells to organize ovarian follicle-like structures (Ohno et al, 1978; Zenzes et al, 1978) means either both mammalian sexes are capable of expressing a pair of ovary-organizing plasma membrane components or ovarian organogenesis is an automatic affair occurring in the absence of H-Y and not requiring specific components. This latter possibility is made unlikely by the observation that avian ovary organization requires the presence of H-Y-like H-W antigen and its specific receptor (Muller et al, 1979b). I believe the key to the testis-organizing role played by the mammalian H-Y antigen-receptor pair is found in H-Y antigen (−) characteristic of male primordial germ cells (see also Figs. 3 and 4).

presence that is suppressing the expression of these two ovary-organizing components by male gonadal cells. Indeed, in an X-linked mutation-caused absence of H-Y antigen, XY wood lemming females readily organized functional ovaries [Fredga et al, 1976; Wachtel et al, 1976]. In this manner, mammalian H-Y antigen came to play the dominance-by-suppression role over ovary-organizing components which may require the dissemination of H-Y antigen by male gonadal cells during the critical stage of testicular organogenesis. Because of this dominant role by testis-organizing H-Y antigen, most of the balanced XX/XY chimeric mice produced by blastocyst fusions succeed in organizing a pair of testes [McLaren, 1976], and the XY-to-XX transfer of H-Y antigen has actually been observed among testicular somatic cells of one such chimeric male mouse

[Ohno et al, 1978b]. In man and marmoset monkeys, early vascular anastomosis between heterosexual twin fetuses causes no ill effect upon sexual development of a female twin [Benirschke and Brownhill, 1963]. In cattle and certain other members of the family Bovidae, however, the same condition invariably retards the ovarian development of female twins, in extreme cases, modifying their gonads to resemble small testes that contain abundant H-Y antigen [Ohno et al, 1976]. A peculiarity of Bovidae may be found in the entrance to blood circulation of testis disseminated H-Y antigen during the normal development of bovine bull fetuses [Wachtel, 1980]. Although an ovary-organizing antigen and its gonad-specific receptor of mammals are not verified entities at the moment, the very fact that XX/XY chimeric mice occasionally succeed in organizing a pair of ovaries [Evans et al, 1977] appears to suggest the existence of H-Y antigen's competitive opposition. Indeed, fetal ovaries of some mammalian species contain an element or elements antagonistic to H-Y antigen [Wachtel and Hall, 1979]. From the above, we see that the inflexibly male-specific and constitutive expression of H-Y antigen is the only new invention that characterizes mammalian gonadal organogenesis, and that this gave testis-organizing H-Y antigen such a pivotal role in the primary (gonadal) sex-determining mechanism of mammals (Table I).

Inasmuch as two or more cell types contribute to the formation of an organ, a set of differentiation plasma membrane antigens that distinguishes individual cell types plays a subsidiary but important role in organogenesis [Boyse and Abbott, 1975]. Here too, reflecting interchangeability of their evolutionary past, mammalian testicular Sertoli cells and ovarian follicular (Granulosa) cells are marked by the same differentiation antigen, instead of each having its own [Ciccarese and Ohno, 1978]. Yet another regulatory economy is thus achieved by conserving the past.

III. ANTIGENIC DETERMINANTS VERSUS RECEPTOR-BINDING ACTIVE SITE OF H-Y ANTIGEN MOLECULE

Characterization of any protein residing on the plasma membrane is a formidable task. We thus resorted to a genetic trick to accomplish the task of identifying H-Y antigen as a discrete protein species. Since H-Y antigen is disseminated by testicular cells [Muller et al, 1979b] to function as a short-range hormone [Ohno, 1976; Ohno et al, 1978a; Ohno et al, 1976], it can not reside on the plasma membrane as an integral component. Instead, it should anchor itself to the plasma membrane via some other components. While the expression of H-Y antigen is ubiquitous in the male, that of specific receptor sites for H-Y antigen is confined to gonadal cells [Ohno, 1976; Muller et al, 1979b; Nagai et al, 1979]. This very fact suggests the requirement for ubiquitously expressed anchorage sites for H-Y antigen. For a number of reasons, I have assigned this anchorage site function to β_2-microglobulin-MHC (H-2 of the mouse, HLA of man) antigen dimers [Ohno, 1977]. Indeed, in the mutational absence of H-Y antigen's proposed anchorage site, β_2m(−), HLA(−) Daudi human male Burkitt lymphoma cells proved incapable of stably maintaining H-Y antigen on their plasma membrane [Beutler et al, 1978; Fellous et al, 1978]. Free H-Y antigen excreted by Daudi cells into the culture medium was identified as a series of polymers formed by interchain disulfide bridges of the quite hydrophobic subunit having a molecular weight between 18,000 and 16,500 [Nagai et al, 1979]. This subunit, with its rather long half-life of roughly 20 hours, readily incorporated ^{3}H-lysine and even more ^{3}H-leucine, but, unique among Daudi excreted proteins, little S-methionine. From a multitude of ^{3}H-lysine-labeled Daudi excreted proteins, gonad-specific H-Y antigen receptor sites residing on the bovine fetal ovarian plasma membrane selectively absorbed only molwt 18,000 × n polymers and quickly reduced them to the monomeric form. On the testicular plasma membrane, endogenous H-Y antigen interfered with this selective uptake by gonad-specific receptor sites. Such an interaction between Daudi excreted human H-Y antigen and its specific receptor site of the bovine origin, if allowed to continue for five days of organ culture, induced the precocious formation of tunica albuginea and seminiferous as well as rete tubules in bovine XX embryonic indifferent gonads [Nagai et al, 1979].

Subsequently, an in vitro mutation affected H-Y antigen excreted by our pseudotetraploid Daudi subline. Its loss of the receptor-binding activity and, therefore, the testis-organizing function, was associated with a mutated subunit's inability to utilize interchain disulfide bridges as the means to polymerize. Instead, because of its increased hydrophobicity, it formed huge aggregates. The integrity of H-Y antigen's receptor binding active site apparently depends upon the presence of one critical cysteine residue. Surprisingly, this mutationally defunct H-Y antigen still retained antigenic determinants and avidly absorbed out the male-specific cytotoxicity contained in H-Y antibody [Iwata et al, 1979].

The above revelation that H-Y antigen serologically so detected may be defunct has an important bearing upon the case of testicular organogenesis failure in an apparent presence of H-Y antigen. In humans, XY women with pure gonadal dysgenesis include the X-linked heritable form [Sternberg et al, 1968] reminiscent of XY wood lemming females [Fredga et al, 1976; Wachtel et al, 1976]. Of 12 such patients examined by Wolf, however, only three did not express H-Y antigen; seven patients, including two sisters, exhibited the normal male level of H-Y antigen [Wolf, 1979]. The etiology of H-Y antigen (+) XY women may be found in mutational defects of the gonad-specific H-Y antigen receptor, as well as in defects in H-Y antigen itself.

IV.   IS AVIAN H-Y (H-W) ANTIGEN OVARY-ORGANIZING?

Just as mammalian H-Y antigen was first found as a male-specific transplantation antigen in the mouse [Eichwald and Silmser, 1955], so was female-specific H-W transplantation antigen found in the chicken [Gilmour, 1967]. The same means also uncovered male-specific H-Y-like antigen in one teleost species[Kallman et al, 1973]. The serological study subsequently suggested the immunological identity between mammalian H-Y antigen and those heterogametic sex-specific transplantation antigens of nonmammalian vertebrates [Wachtel et al, 1975b]. Does the above mean that, when expressed in the heterogametic female sex, the same plasma membrane antigen completely reverses its role in gonadal organogenesis? Since estradiol-induced ovotestis development in ZZ male chicks was found to be associated with the induced appearance in the gonad of H-Y (H-W) antigen, the ovary-organizing capacity of avian H-Y (H-W) antigen can hardly be contested [Muller et al, 1979b]. In a long series of experiments, J. P. Weniger's group has shown that the side-by-side organ culture of a mammalian and avian fetal testicular pair invariably causes ovarian transformation of the latter [Akram and Weniger, 1968]. If the above is due to the mammal-to-avian transfer of H-Y antigen, the same plasma membrane antigen that is testis-organizing in mammals indeed functions as an ovary-organizer in birds. Conversely, however, the mammalian testis in the above experiment may act only as a supplier of androgenic precursors of estradiol-17β. If so, ovarian transformation is due to the endogenously induced avian H-Y (H-W) antigen.

The extreme evolutionary conservation of H-Y antigenic determinants by the heterogametic sex of nearly all vertebrate classes can be explained by the following three alternatives: 1) Antigenic determinants recognized by mammalian H-Y antibody reside outside the functionally critical receptor-binding active site. Our finding on an in vitro mutated human H-Y antigen renders a strong measure of support to this notion [Iwata et al, 1979]. Thus, testis-organizing H-Y and ovary-organizing H-W are different molecules. What then caused the extreme evolutionary conservation of these antigenic determinants? Since its rival antigen responsible for gonadal organogenesis of the homogametic sex tends to be expressed by both sexes, it may have become the general rule of vertebrate evolution that the gonadal organogenesis-directing antigen of the heterogametic sex is to be endowed with the ability to suppress the expression of its rival. This endowment may be embodied in these evolutionary conserved antigenic determinants recognized by mammalian H-Y antibody. 2) Mammalian H-Y and avian H-Y (H-W) and their respective counterparts in reptiles, amphibians, and fish are, in fact, homologous molecules. Whether to function as a testicular organizer or as an ovarian organizer is determined by the type of gonad-specific receptor sites

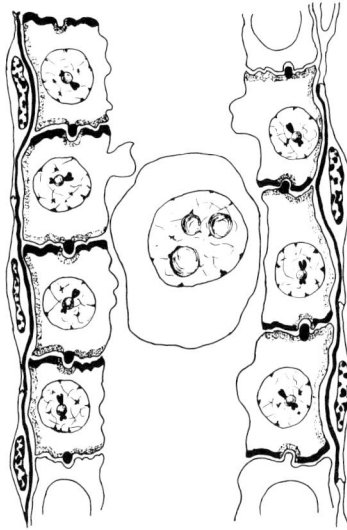

Fig. 1. Schematic illustration of the seminiferous tubule formation by the plasma membrane interaction between mammalian H-Y antigen (solid black) and its specific receptor (shadded). Unusual, polarized distributions of H-Y antigen on the plasma membrane of Sertoli and basement mesenchymal cells contribute to the tubular wall formation [Ohno, 1979]. H-Y antigen (−), receptor (−) primordial germ cells are not directly involved [Zenzes et al, 1978b].

with which each associates. 3) Not only H-Y (H-W) antigen but also its gonad-specific receptor has been conserved throughout vertebrate evolution as the functional pair responsible for gonadal organogenesis of the heterogametic sex. The very characteristic that H-Y (H-W) antigen and/or its specific receptor have been designed to be expressed in germ cells only after they have completed most of the first meiotic prophase stages enables this pair to direct testicular organogenesis as well as ovarian organogenesis. It should be recalled that while male germ cells do not enter meiosis until puberty, female germ cells complete up to the diplotene stage of the first meiotic prophase during the act of ovarian organogenesis. Befitting their presumptive derivation from mesonephric duct epithelia, the tendency to form a tubular structure is inherent in the gonadal somatic elements that can differentiate into either testicular Sertoli cells or ovarian follicular (granulosa) cells. Indeed, in early differentiating gonads, one finds female counterparts of seminiferous tubules in ovigerous cords stuffed with primordial germ cells radiating from the future rete region toward the periphery [Gropp and Ohno, 1966] Their obscurity in relation to male seminiferous tublues is found in extremely expanded diameters of these ovigerous cords at the periphery as well as to their short, stubbiness. As to how the plasma membrane interaction between H-Y antigen and its specific receptor causes the seminiferous tubule formation in mammalian males, our histochemical study revealed the distinctive bipolar distribution of this antigen on the Sertoli plasma membrane and its equally unique unipolar distribution on the surface of basement membrane mesenchymal cells as schematically illustrated in Figure 1 [Ohno, 1979]. The tubular wall formation must be due to the complementarily bipolar distribution of H-Y receptor sites on the Sertoli plasma membrane. Female ovigerous cords must also be formed by an analogous plasma membrane interaction. It has been observed that mammalian testicular germ cells are endowed with neither H-Y antigen nor its specific receptor until puberty [Zenzes et al, 1978b]. It follows then that the presence of male primordial germ cells does not disturb the seminiferous tubular wall structure maintained by strong plasma membrane interactions between Sertoli cells alone. Ovigerous cords of avian female gonads are far more conspicuous than their mammalian counterparts, their deeper portions not penetrated by germ cells becoming long and tortuous tubules of the very prominent medulla. The above may indeed indicate the involvement of H-Y antigen and its receptor in the avian ovary organogenesis. As primordial germ cells first differentiate into oogonia and then to oocytes, and as these oocytes complete up to diplotene stage of meiosis, dictyate oocytes now would express H-Y antigen and/or its specific receptor. Nearby epithelial elements of the ovigerous cord wall would now be attracted as much toward the plasma membrane of individual dictyate oocytes as to each other; the formation of primordial ovarian follicles follows (Fig. 2). I find this third alternative particularly attractive. Its one shortcoming, however, lies in an implied complementary behavior of a rival pair of plasma membrane components responsible for gonadal organogenesis of the homogametic sex.

V. WHENCE COME TRUE HERMAPHRODITES?

As the induction by estradiol-17β of ovotestes in ZZ male chicks was found to be associated with the induced appearance of H-Y (H-W) antigen in their gonads [Muller et al, 1979b], complete sex reversals caused by environmental manipulations in fish [Yamamoto, 1961] and amphibians [Mikamo and Witschi, 1964] are also likely due to the suppression in the heterogametic sex and the induction in the homogametic sex of H-Y-like antigen expression. The situation found in turtles in which the sex is determined by incubation temperatures of hatching eggs [Pieau, 1975] can also be explained by the acquired temperature dependence in H-Y-like antigen's expression. Their chromosomal sex-determining mechanism needs not have disappeared. Both protoandrous (males when young) and protogynous (female when young) types of asynchronous herma-

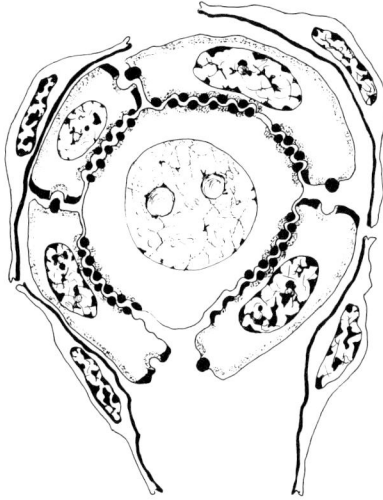

Fig. 2. The scheduled expression of H-Y antigen and its specific receptor on the plasma membrane of dictyate oocytes may enable avian species to utilize the set of plasma membrane components that is testis-organizing in mammals for their ovarian organogenesis.

phroditism rather frequently encountered among teleost fish [Chan, 1970] can also be explained by the age-dependent sequential induction and suppression, or vice versa, of H-Y-like antigen's expression. Thus, they may have evolved from gonochorist ancestors in which either male or female heterogamety operated.

As to evolution of synchronous hermaphrodites having been endowed with ovotestes or testes and ovaries, the existing threshold level in H-Y-like antigen's capacity to direct gonadal organogenesis has to be realized. In mammals, the XX male condition due to somewhat subnormal expressions of H-Y antigen in an apparent absence of the Y and the condition of XX true hermaphroditism appear to represent two sides of the same coin. For example, an H-Y antigen (+) bitch who gave birth to XX male puppies was herself endowed with a pair of ovotestes, although this XX male condition was thereafter inherited as an autosomal dominant trait [Selden et al, 1978]. Furthermore, implicit in the finding that the XX male condition may also be inherited as an autosomal recessive trait, as in the goat [Wachtel et al, 1978] and man [de la Chapelle et al, 1978], is the fact that there exists a threshold in the H-Y antigen level below which this antigen has no influence on gonadal organogenesis, for mothers of these XX males are obligatory heterozygotes expressing, at the most, half as much H-Y antigen as their XX sons. It follows then that a certain range of H-Y antigen levels somewhat above the threshold must necessarily cause a stalemate in the testicular versus ovarian organogenesis decision of embryonic indifferent gonads. We then expect the constitutive but substantially subnormal expression of H-Y-like antigen in synchronously hermaphroditic species, which are also rather common among teleost fish [Chan, 1970].

In view of the antiquity of the chromosomal sex-determining mechanism, it is really no surprise to find the involvement of evolutionary conserved H-Y plasma membrane antigen in the testis versus ovary decision of all vertebrate gonads. Although birds are already equipped with a conspicuously heteromorphic sex chromosome pair, the expression of H-Y (H-W) antigen can be induced in the homogametic ZZ sex [Muller et al, 1979b], thus, ruling out the W-linkage of this ancient gene. The Y-linkage of mammalian H-Y structural gene might also be questioned.

ACKNOWLEDGMENT

This work was supported by National Institutes of Health grants.

REFERENCES

Akram H, Weniger JP (1968). Arch Anat Microsk Morphol Exp 57:369–378.
Allen CE (1917). Science 46:466–467.
Benirschke K, Brownhill LE (1963). Cytogenetics 2:331–341.
Beutler B, Nagai Y, Ohno S, Klein G, Shapiro I (1978). Cell 13:509–513.
Boyse EA, Abbott J (1975). Fed Proc 34:24–27.
Chan STH (1970). Phil Trans R Soc Lond B 259:59–71.
Chapelle A, de la, Koo GC, Wachtel SS (1978). Cell 15:837–842.
Ciccarese S, Ohno S (1978). Cell 13:643–650.
Ebeling AW, Chen TR (1970). Trans Am Fish Soc 99:131–138.
Eichwald EJ, Silmser CR (1955). Transplant Bull 2:148–149.
Evans EP, Ford CE, Lyon MF (1977). Nature 267:430–431.
Fellous M, Gunther E, Kemler R, Wiels J, Berger R, Guenet JL, Jakob H, Jacob F (1978). J Exp Med 148:58–70.
Fredga K, Gropp A, Winking H, Frank F (1976). Nature 261:255–257.
Gilmour DG (1967). Transplantation 5:609–706.
Gordon M (1951). Zoologica 32:27–134.
Gropp A, Ohno S (1966). Z Zellforsch 74:505–528.
Iwata H, Nagai Y, Stapleton DD, Smith RC, Ohno S (1979). Arthritis Rheum 22:1211–1216.
Kallman KD, Schreibman MP, Brokoski V (1973). Science 181:678–680.
Krco CJ, Goldberg EH (1976). Science 193:1134–1135.
Lyon MF (1961). Nature 190:372–373.
McLaren A (1976). "Mammalian Chimaeras, Development and Cell Biology," 4. Cambridge University Press, London.

Mikamo K, Witschi E (1964). Experientia 20:622–624.

Miller RA (1938). Anat Rec 70:155–189.

Moscona A (1957). Proc Nat Acad Sci USA 43:184–189.

Muller U, Wolf U, Siebers JW, Gunther E (1979a). Cell 17:331–336.

Muller U, Zenzes MT, Wolf U, Engel W, Weniger JP (1979b). Nature 280:142–144.

Nagai Y, Ciccarese S, Ohno S (1979). Differentiation 13:155–164.

Ohno S (1976). Cell 7:315–321.

Ohno S (1977). Immunol Rev 33:59–69.

Ohno S (1979). Major sex determining genes. In "Endocrinology Monograph Series." vol 11. Springer-Verlag, Berlin.

Ohno S, Christian LC, Wachtel SS, Koo GC (1976). Nature 261:597–598.

Ohno S, Ciccarese S, Nagai Y, Wachtel SS (1978a). Arch Androl 1:103–109.

Ohno S, Nagai Y, Ciccarese S (1978b). Cytogenet Cell Genet 20:351–364.

Pieau C (1975). In "Intersexuality in the Animal Kingdom." (R Reinboth, ed.), pp 332–339. Springer-Verlag, Berlin.

Selden JR, Wachtel SS, Koo GC, Haskins ME, Patterson DF (1978). Science 201:644–646.

Sternberg WH, Barclay DL, Klowpfer WH (1968). N Engl J Med 278:695–770.

Wachtel SS, Basrur P, Koo GC (1978). Cell 15:279–281.

Wachtel SS, Hall JL (1979). Cell 17:327–330.

Wachtel SS, Koo GC, Boyse EA (1975a). Nature 254:270–272.

Wachtel SS, Ohno S, Koo GC, Boyse EA (1975b). Nature 257:235–236.

Wachtel SS, Koo GC, Ohno S, Gropp A, Dev VG, Tantravahi R, Miller DA, Miller OJ (1976). Nature 264:638–639.

Wachtel SS, Hall JL, Müller V, Chaganti RSK (1980). Cell 21:917–926.

Wolff Et, Ginglinger A (1935). Arch Anat Strasbourg 20:219–278.

Wolf U (1979). Hum Genet 47:269–277.

Yamamoto T (1961). J Exp Zool 146:163–180.

Zenzes MT, Muller U, Aschmoneit I, Wolf U (1978a). Hum Genet 45:297–303.

Zenzes MT, Wolf U, Gunther E, Engel W (1978b). Cytogenet Cell Genet 20:365–372.

Subject Index

A

3-Acetylpyridine. *See* NAD

*ts*34-AEV cells
transformation of mesenchymal cells with, 160, 164–168

wt-AEV cells
transformation of mesenchymal cells with, 160, 164–168

Alu family sequence
repeated DNA, 197, 198

Amiloride
inhibition and, 176
sodium transport and, 173–176

6-Aminonicotinamide. *See* NAD

Antheraea polyphemus, 69, 70

Antibodies
use of, 21

Aspergillis
mitochondrial DNA of, 121

ATPase
differentiation and, 172
polypeptides and, 120

Autosomal genes
evidence for, 226

Auxin
action of, 83–84, 100
basal hypocotyl, effects on, 104–109
excised elongating hypocotyls, effects on, 109–116
gene expression and, 90–94
gene regulation and, 83–96
polyadenylated RNA, regulation of, 85–95
protein synthesis and, 105–109, 111–117

soybean hypocotyl, effects on, 100–104, 112

Auxotrophs
nutritional-type mutants and, 145–147

B

B-A chromosome translocations
mutants, analysis of, 152

Basal hypocotyl
auxin, effects on, 104–109
protein synthesis of, 105, 116–117

Bombyx mori, 71

Bovine freemartin
H-Y antigen, role of, in, 229

N-Butyrate
as an inducer of differentiation, 167–168

C

Calcium flux
commitment, relationship to, 173–176, 180

Cartilage
development, 18–19, 44, 46
differentiation, 19–20
Meckel's, 19, 22, 28
molecular analysis of, 22–24, 28–29
nanomelic mutant, defects in, 24–25
proteoglycans, 20–22, 23, 55
sternal, 22, 26, 28
ultrastructural features of, 26

Cell
development response, 66
physiological response, 66

F

Fertility
  reversion of *cms*-s to, 133–134
Fetal genotype, 205
Fetal nutrition
  maternal environment and, 202, 206
Fetal size
  maternal size and, 203, 204
Flow cytometry
  membrane potential and, 177–179
Friend virus complex
  murine erythroleukemia cells and,
    160
Frizzle fowl, 6–8

G

Gene duplications
  DNA exchanges between, 192–193
  evolution of, 189
  general aspects of, 189
Gene expression
  auxin-regulated, 90–94
  DNA sequence, changes in, 158
  experimental results of, 63–64
  influence of extracellular matrix on,
    58–64
  globin, in chick embryogenesis,
    157–169
  molecular heterogeneity and,
    127–128
401/18 Gene pairs
  organization of, 78–80
Gene regulation
  auxin and, 83–96
  rRNA of, by auxin, 84–85
  structural, auxin and, 85–88, 93–94
Genes
  autosomal, evidence for, 226
  modifiers, 10–11
  normal action of, 10–11
  X-linked, evidence for, 225–226
  Y-linked, evidence for, 225–226
Gene structure
  chorion, of, 71–78

Genetics
  development, 1–11
  H-Y antigen, of, 225–227
Globin genes
  anatomy of, at the DNA level,
    186–187
  co-evolution and control of, 185–199
  expression of, in chick
    embryogenesis, 157–169
  human fetal, duplication of, 190–192
  pseudogenes in, 193–198
α Globin gene clusters
  anatomy of, 186–187
β globin gene clusters
  anatomy of, 186–187
  evolution of, 190–191
  repeated sequences in, 198–199
Golgi complex
  cytological characterization and, 29
Growth, compensatory
  post-implantation embryo, in,
    208–214
Growth, control
  size of organisms and, 202, 207

H

hCG receptors
  xx converted gonadal cells in, 225
Heterogametic gonad
  induction of, 219–232
Hogness box, 75
Human fetal globin genes
  duplication of, 190–192
H-W antigen. *See* H-Y antigen
H-Y anchorage site, 228, 240
H-Y antigen
  antigenic determinants versus
    receptor-binding active site of,
    239–240
  avian, 241–243
  bovine freemartin and, 229
  definition of, 219–221
  discovery of, 227
  expression of, 227

DATE DUE

DEMCO 38-297